Common Sense Cosmology

By

Lenard Metzger

Elemental Publishing
Rochester, New York

COMMON SENSE COSMOLOGY

Elemental Publishing
80 Westerloe Avenue
Rochester, NY 14620
(585) 473-9303

Library of Congress
Control Number: 2010930766

ID: 332616
lulu.com

ISBN: 978-0-557-51848-7

PREFACE

I have been trying to decide how to present the ideas that I have been thinking about for a long time. What I have decided to do is to use a narrative style rather than a textbook style. There is the question of what to call these ideas. They certainly do not have the status of "theories". "Hypothesis" is much too fancy a word. I am not sure about "conjectures". I will call these concepts or ideas and leave off the term "crackpot".

At about my ninth or tenth birthday, my sister gave me a book with a title something like "The Wonders of Science". I learned many wonderful things reading it. All the strange things people have believed, over the years, about the world and the universe, puzzled me.

How could people believe that the world was flat and that if you sailed a ship to the edge your ship would fall off? Why didn't they realize that all the water in the seas would pour off the edges and the ship would have nothing to sail on?

After it was generally accepted that the world was round, there was this idea that the Sun and the stars all revolved around the Earth. And even when it was finally understood that the Earth revolved around the Sun, the belief was that the stars were all on the surface of a crystalline sphere that rotated around us.

The thought then was that the region beyond this was heaven, the abode of the angels. Then people realized that some of the "stars" moved erratically, along different paths then those of the rest of the stars.

These objects were named "planets" and were thought to be on separate moving crystalline spheres.

It was only later that true astronomers described the real situation, that the Earth is one of a group of planets circling the Sun and that the stars are at a very much greater distance away.

How could people have believed in crystalline spheres with nothing beyond them? In my teens I tried to visualize this but every time I reached a mental barrier in space I continued outward from the other side of the barrier. I soon concluded that there was no end to space.

When I first learned about what was thought to be the structure of atoms in matter, I was struck by how similar their structure was to that of our solar system, with the electrons circling the atomic nucleus, just as the planets circle the Sun. When I learned of galaxies and molecules, I thought how their relationships were similar to that of solar systems and atoms. Later, I thought, what if galaxies are organized similar to the way molecules are assembled into crystals? The final leap was to imagine the entire universe as a small object in some much larger scheme of things. I painted a picture of this concept and later wrote a poem about it.

Going back to the beginning, when I was around seven or eight years old, I was fascinated by the ham radio that my older brother had built. He got tired of my playing with it and unplugging the coils and putting them in the wrong sockets. So he bought a crystal set for me and helped me build my own little radio.

Later I looked at his copies of "Popular Mechanics" and "Radio and Electronics". While in high school, I started reading science fiction magazines, such as "Fantastic" and "Amazing". I progressed to more scientific kinds, such as "Astounding" and later "Analog". At that time I started reading "Scientific American" and have persisted with it to this day.

I had developed an interest in nuclear physics from these sources and when I started college I took courses in that field. I took a several year break for military service, where I became a radio operator.

After WW-II, I went back to college, but changed my specialization to electronics. I figured, with the "Bomb" developed, that nuclear physics would not be as interesting. I did take some courses in this area, such as nuclear instrumentation and atomic and molecular spectroscopy.

My working career was spent doing systems development work, mostly in electronics, but with a smattering of optics and mechanics. My early work was on classified projects that didn't require patent applications. In my later work developing consumer products, I received dozens of patents.

Since my retirement, I have had more time to think about really important things. I have gone back to my interest in cosmology, rereading my collection of books on relativity. Through the years of reading "Scientific American" I have watched the passage of the various theories on physics and astronomy.

Several years ago some articles triggered my desire to respond to what I considered a lack of logic in their premises. A couple of the articles made various uses of what Einstein considered to be his greatest mistake, the arbitrary coefficient that he introduced to make space not expand.

Another pair of articles appeared on adjacent pages of an issue. One article considered light as an electromagnetic wave; the other one treated light as photons imparting momentum. It struck me that there ought to be a way of combining these concepts. I will describe my thought on this later.

Now it is about time for me to start disclosing my thoughts about what I think the universe is all about. I will start at the present time, going from the smallest objects to the largest and then go back to the beginning, and then forward to the ending and back to the beginning.

Worlds Within Worlds
By
Lenard Metzger
Circa 1945

RELATIVITY
(Haiku for Albert)

We don`t see black holes
Because of their gravity
Light is too heavy

Our cosmic space curves
Warped by all the galaxies
Nothing can escape

Someone very far
Beyond our black universe
Won`t see us in here

By
Lenard Metzger
Circa 1985

Contents   Pages

INTRODUCTION

What do I mean by common sense cosmology? I believe that common sense should be used as the criteria in deciding between possible explanations about anything. I consider cosmology to be the study of everything, including our universe and its place in the cosmos. We will consider our universe, as it appears to exist at this time, and then look at where it came from and where it is going. To be able to do this we will have to establish some fundamental assumptions.

I believe that it is safe to assume that our universe began with what is called the "Big Bang". An expanding universe was postulated based on the red shift of various stars and galaxies, as a function of their distance from the Earth. Hubble and others quantified this relationship. The age of the universe was estimated at between 10 and 20 billion years. This was arrived at by projecting the estimated rate of expansion (the Hubble Coefficient) backward in time to where all the observed bodies would have coincided.

The measurement of the microwave background radiation of space and its equivalent black body temperature tended to verify the time required in cooling the initial, extremely high temperature to that which is measured now.

According to Einstein's theory, if the expanding universe had above a certain average density of matter within it, the universe would be what is deemed "closed".

The universe would eventually stop expanding and start to collapse back toward a common center. If the average density were appreciably less, the universe would be considered "open" and would keep expanding forever.

I believe that the latest estimate of the total visible and probable dark matter in the universe is less than the required amount to close the universe. It was hoped that the universe would be closed and that eventually all the matter would collapse back to the center and perhaps result in a new "Big Bang".

I believe it was suggested that this would take about a trillion years and that all of the stars would have died out. Most would have been absorbed into the central black holes of the galaxies. The remainder would be scattered individual black holes and neutron stars. Now, it is assumed that the expansion of the universe will continue without end.

Considerable theoretical work has been done on the earliest time of the universe. The present theory is that the explosion's temperature was so high initially, that only gravitational energy existed.

As the expansion continued and the temperature dropped, successive energy régimes occurred, strong force, weak force and electro-magnetic force. Soon, subatomic particles appeared, followed by the first atoms, hydrogen, deuterium and helium. Some postulate that a period of greatly increased rate of expansion occurred earlier.

As time went on the rapidly expanding sphere of hot gas changed from a homogeneous sphere into one with spaces and voids.

Over vast periods of time, the hydrogen gas converged, becoming numerous stars that combined into galaxies. Studies of the distribution of galaxies as a function of distance from us indicate that space is essentially flat.

The above concepts will be the starting point for my new ideas. The approach that will be taken is to use the least amount of mathematics possible and use simple logic to describe my vision of how the cosmos might work. Generally accepted astronomical and physical facts will be assumed as givens. Conclusions that have been reached as the result of manipulations of derived equations will be questioned and subject to re-interpretation. There are a number of prevalent theories that are not so easily understood or visualized. In some cases they are not even logical. In the following sections I will address these cases in detail and offer alternative explanations.

STARLIGHT

To get yourself into the proper mood to understand cosmology, imagine you are outside, somewhere far from the lights of your town. The moon has not risen into view. Look up at the dark sky full of stars. Your feet are on the ground. You feel the whole Earth under you. You know that it is a huge ball, so big that people used to think that it was flat. If you stand and watch the stars you will realize that they are slowly moving across your view. But it is you that is moving as the Earth rotates with you being taken along.

The sky is dark because the Sun is shining on the other side of the Earth. You wont see the Sun again until the Earth makes another half turn. I am sure that you understand that the Earth also revolves around the Sun, taking about 365 days to go all the way around once. This we call a year. The ancient name for the Sun was Sol. That is why we call things about the Sun, solar.

The Sun is our closest star. It is only about 93 million miles away. This may seem far away, but compared to the distances to the other stars it is not far at all. We do not usually talk about the distance to a star by using miles. Instead, we use the speed of light and how long it takes for light to travel from one place to another as a measure of distance.

It takes light about 8 minutes to travel from the Sun to the Earth. For the light of the next closest star to reach us takes over 4 years.

Perhaps you can see a narrow white band crossing the sky. This is called the Milky Way. The Sun and all the stars you see with your naked eyes are part of what is called the Milky Way Galaxy. Our galaxy is composed of billions of stars arranged in a flattened disc shape. Our Sun is located about half way between the center and the edge of the galaxy.

When we on Earth look toward the center of the galaxy, the stars appear to get closer and closer together, until they seem to become a solid band of white, the Milky Way. If we were to look between the stars through a large telescope into the far reaches of the universe we would see many other galaxies like the Milky Way. There may be billions of them.

Almost everything we see in the sky is radiation from hot hydrogen atoms in the outer regions of stars. A very small amount of the light comes from the other trace elements. To understand how this happens we must look at the hydrogen atom.

The electron spins around the nucleus (a proton) at a relatively large distance, compared to the size of the nucleus. The atoms in the gas interact with each other more frequently with increased temperature. This raises the energy of the electrons in some of the atoms. The electrons return to their normal energy levels by emitting photons of light.

It was found that the photons emitted from hydrogen atoms had a specific pattern of discrete wavelengths, as measured by a spectrometer. The radiation from other elements had their own distinctive patterns.

The measurement of the wavelength of light is called spectroscopy and has been very important in the understanding of stars, galaxies and the cosmos. Measuring the spectra of the light emitted has led to the knowledge of star size, temperature, and direction and speed of motion. The hotter the star the higher the energy of the excited hydrogen in its atmosphere and therefore the bluer its radiated light due to the higher energy photons.

Color is an important characteristic of light. It is what we perceive in the visible portion of the spectrum of light. But we should consider it as the wavelength of light, whether it is infrared, at a longer wavelength than visible light, or ultra-violet light, at a shorter wavelength than visible light. As an electro-magnetic wave, a specific "color" of light oscillates at a specific frequency as it travels along at light speed. The distance that it travels in one cycle of oscillation is the wavelength.

As an example, the wavelength of red light is almost a micron, which is a millionth of a meter. Since the speed of light is about 300 million meters per second and the frequency is the speed divided by the wavelength, this gives a frequency for red light of about 300 trillion cycles per second. This is why light is usually characterized by its wavelength.

An important phenomenon to understand is what is called the "red shift". This is used to measure the velocity of a star or galaxy relative to the Earth. The greater the velocity of separation, the more the characteristic blue light appears to be shifted towards the red end of the spectrum.

To understand this, consider a star moving away from us at a tenth the speed of light. Think of the first cycle of the light at the beginning of one second, moving the known distance of 300 million meters during that time. The last cycle, emitted after the star had traveled one second, would be 330 million meters behind the first cycle. Dividing this slightly greater distance by the known frequency of the light results in a slightly longer wavelength. Therefore, when we detect this radiation, it appears shifted towards the red end of the spectrum.

Another aspect of this is that the same effect would occur if the distant object were stationary and we were moving away from it at some fraction of the speed of light.

During a one second period, that we would measure the light, slightly less of the radiated wave would pass us. This would result in a lower number of cycles per second being detected. This results in a longer wavelength, redder light observed.

This is consistent with the Theory of Relativity assumption that only relative velocity can be determined and that the speed of light is invariant. The speed of light is an important factor in the study of cosmology. The theory of relativity established the relationship between mass and energy and its dependence on the speed of light.

In Figure 1, the path of a free electron is shown as it approaches a positively charged hydrogen ion. As it nears the proton, the electron emits a photon of light and goes into orbit around the nucleus.

If this orbit happened to be the closest one to the proton that can occur, the photon emitted would have the largest momentum of any of the hydrogen photons in the hydrogen spectrum. It would have a frequency that placed it in the ultra-violet region.

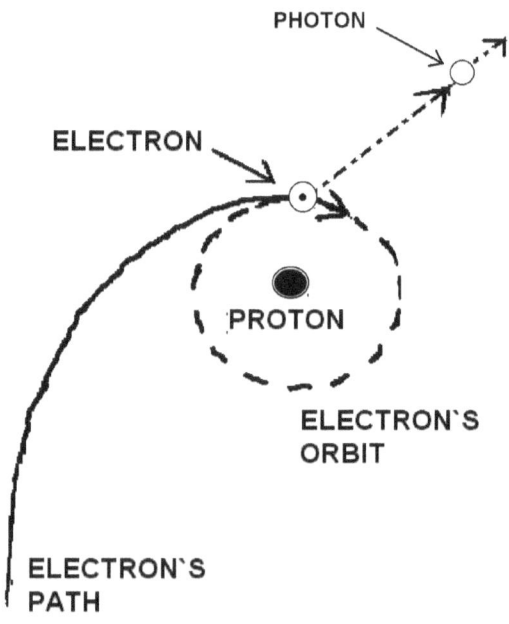

Figure 1

A theory about the hydrogen atom has been named The Bohr Model after the scientist who developed it. A representation of this is shown in Figure 2. This model has a number of nested orbits with systematically increasing radii. His theory states that electrons could only enter orbits with specific energy levels. The photons emitted would have momentum determined by the difference in energy levels between the starting and ending orbits.

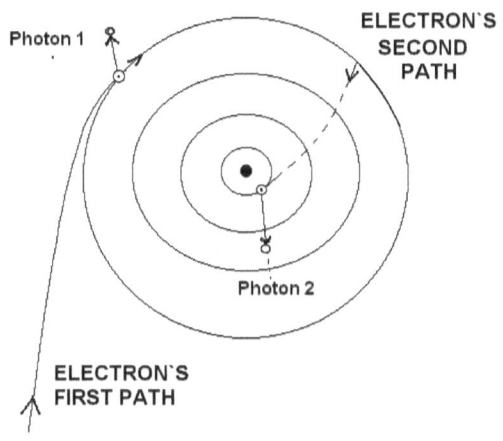

Photon 1

ELECTRON'S
SECOND
PATH

Photon 2

ELECTRON'S
FIRST PATH

Figure 2

In Figure 2 the free electron first entered one of the outer orbits and the energy difference was small. Therefore, the first photon was of low momentum and frequency. The second transition shown was into the innermost orbit and had a larger energy change.

The second photon had a larger momentum and higher frequency than the first. Interestingly, the sum of these two photon energies equaled the energy of the photon shown in Figure 1.

The electron transitions from the various orbits going directly to the innermost orbit produce a spectrum with numerically related frequencies. This spectrum is called the Lyman spectrum after the scientist who discovered it.

Similarly, the transitions that end at the next closest orbit to the nucleus produce a spectrum named after its discoverer, Balmer. The highest frequency photons in this series are in the blue end of the visible spectrum.

All of these and other spectra were mixed together in the light coming from the Sun. The early scientists studying them sorted them out with some difficulty. Laboratory tests under controlled conditions helped. After establishing the nature of these spectra they were able to detect the shift that had occurred in the spectra from the light of distant stars and galaxies.

It should be noted that there are two kinds of hydrogen spectra. One consists of the bright lines emitted by the excited hydrogen atoms. The other spectrum consists of dark lines in the almost continuous bright light from the star. This spectrum is produced when cooler hydrogen atoms, selectively absorb photons passing through the outer region of the star.

For a further discussion of my ideas about photons, see Addendum A. I used Planck`s equation for the energy of a photon and Einstein`s equation for the equivalence of mass and energy to derive an equation for the mass of a photon. The conclusion I reach is that a photon has a mass that is a very small fraction of the mass of an electron. For this to be true, a photon must travel at a speed slightly less than the theoretical, limiting speed of light, "c".

STARS

How does hydrogen, the smallest element, create stars? An electron spins around a single proton, the nucleus of the hydrogen atom. A proton is made up of even smaller components called quarks. It contains two "Up" quarks, each with a charge of +2/3. It also contains one "Down" quark with a charge of −1/3. This is shown in Figure 3. It shows how a very high-energy electron can interact with a hydrogen ion (a proton) and produce a neutron.

The electron can be thought of as having combined with one of the Up quarks, changing it to a Down quark. The resulting particle then contains two Down quarks and one Up quark. This is a neutron.

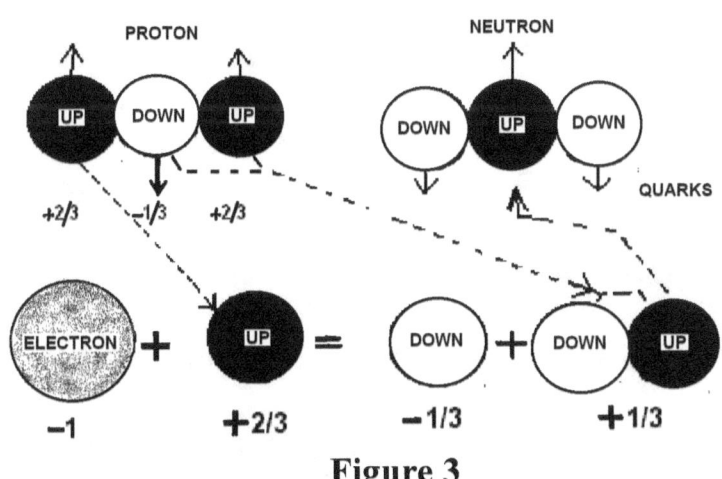

Figure 3

When a neutron combines with the nucleus of a hydrogen atom, it forms the isotope of hydrogen called deuterium. For more information about these atoms, see Addendum C.

These two atoms, hydrogen and deuterium, were the first ones that formed, after the start of our universe. They constituted the greatest majority of the matter in the universe. However, the amount of hydrogen was far greater than that of deuterium.

These atoms formed gasses that gathered into large, globular concentrations. This began the process of star formation. The weight of a globular volume of gas produced a high pressure at the center. It was the presence of the deuterium at the center of the star that allowed an easier start of nuclear fusion. Two deuterium atoms combined to form an atom of helium, which has a slightly smaller mass than two deuterium atoms. The net loss of mass turned into energetic, subatomic particles, which heated the core of the star. The increased heat exerted enough pressure to resist the inward pressure of the great mass of the star.

Over an extended period of time, the supply of deuterium became depleted and the core temperature started to decrease. The weight of the star would compress the core further until the next stage of nuclear fusion started. Helium atoms started to form heavier atoms, with the generation of enough additional heat to stop further compression. This sequence would repeat until the final fusions produced iron atoms.

This was the beginning of the end for the star, since iron would not support this kind of fusion. Many of these first generation stars would eventually explode and disperse the various elements into space where they would mix with the hydrogen clouds. From these clouds the second-generation stars were born.

In these stars the nuclear process in the core produced neutrons that bombarded the heavier elements, such as iron. This produced even heavier elements over time.

Our Sun is a second-generation star. It is one of the smaller sized stars. It is used to compare the size of other stars. We call our Sun "one solar mass". The smallest stars are about one half a solar mass and the largest are more than one hundred solar masses in size. The amount of nuclear fusion needed to withstand the weight of a star varies with the size of the star. Our Sun has enough fuel to burn for billions of years. The larger stars may have to use up all their fuel in less than a million years. When a star has used up its fuel, a number of different things can happen. Some stars become large cool red stars. Some become small hot dwarf stars. Stars that are in the range of ten solar masses can explode and then collapse, becoming a neutron star.

The most spectacular event occurs when a star of about one hundred solar masses explodes as a super nova. It may then collapse into a black hole.

BLACK HOLES

I have been studying the current theories of the formation of neutron stars and black holes. Mathematicians have done much of the theoretical work on black holes. Their equations for the density of matter at the end of the collapse into a black hole contain a term in the denominator that goes to zero and causes the equation to go to infinity. Hence, a black hole with an infinite mass density and zero size at the center is predicted. However, they are said to have a finite mass. This is a singularity.

The nature of neutron stars may be the best place to start in understanding my idea about black holes. I read that neutron stars are thought to have a surface shell of iron atoms, which are the remnants of the depleted nuclear core. Inside of the surface shell are the neutrons, essentially in contact with each other. In the collapse of the core the electrons of the heavier atoms are forced into the nuclei. The electrons and protons combine, forming neutrons, as shown in Figure 3.

I also read, that at the center of the neutron star there could be some exotic particles. I started wondering what these exotic particles could be. This led me to the idea of neutral particles, made up the next heavier quarks, becoming the predominant component of the individual black hole from a collapsed star.

There should still be the layered structure of an outside layer of iron atoms. Within that there should be a layer of neutrons and inside of that there would be the particles made up of the charm and strange quarks. I really need a shorter name for those particles, such as "charstrons", perhaps.

Neutron stars are said to result from stars of about eight solar masses. After they collapsed, they had masses of a little over one solar mass. I am assuming that this was essentially the mass of their nuclear core. The remainder of their mass must have been mostly hydrogen gas blown off in the explosion. I understand that these bodies ended up with a diameter of about ten to twenty miles. In Figure 4 the relative size of different mass stars is shown. Their masses increase with their volumes, and their volumes increase as a function of their diameters cubed.

STAR SOLAR MASSES

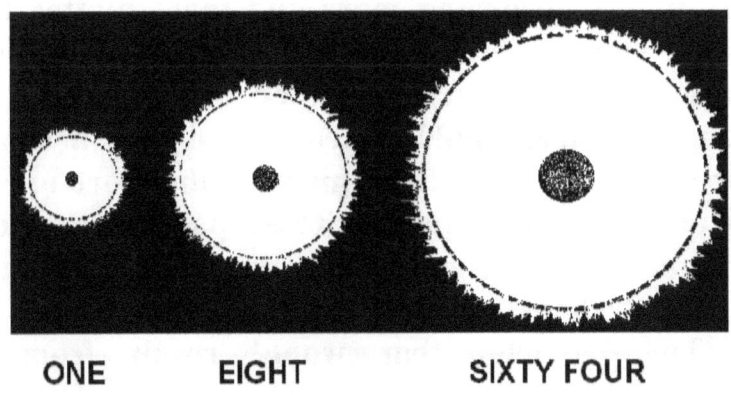

ONE   EIGHT   SIXTY FOUR

Figure 4

Black holes are supposed to occur some of the time when larger stars, in the range of fifty to one hundred solar masses, collapse into super novas. I am assuming that a sixty-four solar mass star would have a core of about eight solar masses.

If it had collapsed into a neutron star, which it might in a transient phase, it would have had a diameter twice that of the above neutron star. As the collapse continued, most of the neutrons would be converted into a fewer number of more massive particles; and the diameter would continue to shrink. If the ratio of neutrons to the "charstrons" were about one hundred to one, the diameter would be reduced to about five to ten miles. It should still have about the same mass of eight solar masses, but squeezed into a much greater density. For this mass at that radius, light would not be able to escape the effect of its gravity.

If this first kind of black hole were in a situation where it was absorbing more and more matter from adjacent stars, it might reach a state where it would collapse again. A large number of "charm" and "strange" quarks could be converted into a smaller number of neutral particles made of the more massive "top" and "bottom" quarks. These particles would be about one hundred times as massive as the "charstron" particles. Perhaps, we could call them "tobotrons".

The particles that would result from the conversion of quarks of one type to those of another are shown in Figure 5.

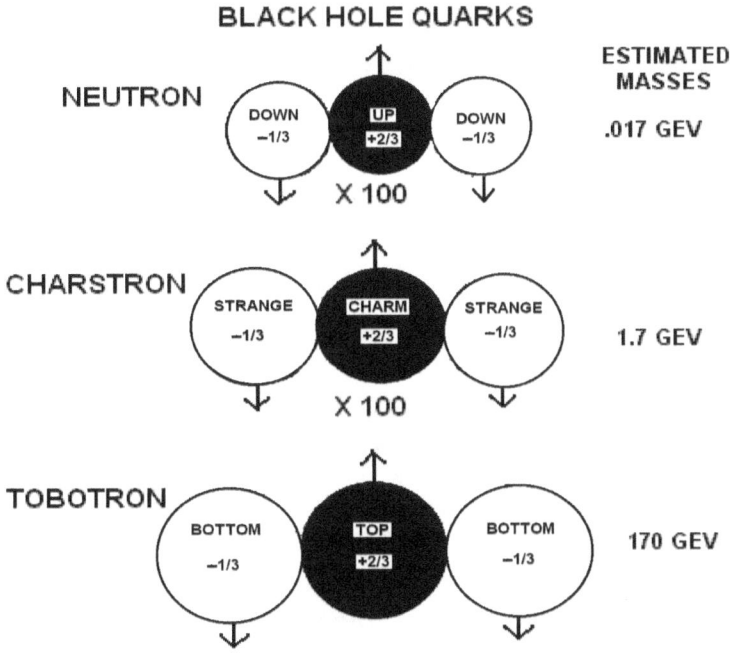

Figure 5

There is evidence that there are massive black holes at the center of many galaxies. They appear to have masses in the order of millions of times that of our Sun. This suggests to me that these are black holes of the second kind, or possibly a third kind.

If enough of these black holes combine, eventually a point will probably be reached where the exotic particles of the core can no longer withstand the gravitational pressure and they will implode and turn into an explosion of gravitational energy. This would be somewhat like what we call the "Big Bang".

BIG BANG

A recent article described observations of quasars, estimated as emitting light towards us about 12 billion years ago. This was thought to have taken place only one billion years after the Big Bang. Repeated observations of these objects indicated that the rate of change of their red shift was increasing. It was concluded that the expansion of the universe was accelerating. A negative force of gravity was postulated to account for this occurrence, using Einstein's infamous variable constant.

Studies of the distribution of galaxies as a function of distance indicate that space is essentially flat. I will take a Newtonian view of the universe and assume space is linear and not expanding. I believe there is a different way of explaining an increasing red shift in the above example. The object may have been moving very slowly, either toward us or away. If it had a decelerating outward motion or an accelerating inward motion, the result would be an increasing rate of red shift change.

Instead of the usual assumption that all galaxies are moving away from us with velocities that increase with distance, assume that astronomical bodies move away from where the Big Bang occurred with a wide range of radial velocities. Also, assume that the Milky Way galaxy has about an average radial velocity.

The amount of red shift of light we observe from distant objects depends on the rate of increase of the distance between them and our galaxy.

There are several ways that we may observe the same degree of red shift. The distant source of light could be moving in a direction opposite to the motion of our galaxy. The observed red shift would be a function of the sum of their velocities.

There could be an object moving in the same direction as our galaxy, but much slower. It would then be closer to the center of the universe than our galaxy. Another object could have started in the same direction as our galaxy but with a greater radial velocity. It would then be farther away from the center than our galaxy. In these cases the red shift would be directly proportional to the difference in velocities between them and us.

Figure 6 illustrates these things. Assume that our galaxy is located at point A. Its radial velocity is shown resolved into vertical and horizontal components. Point A is shown at a distance from the center (The Big Bang) that is near the peak of the galaxy population density

A galaxy at point B had started with a larger radial velocity than our galaxy. The vertical velocity component of the B galaxy was greater than that of our galaxy at the time that B emitted the light that we now receive. The red shift that we observe is a result of the difference between the two vertical velocities.

. The galaxy at point D started out with a smaller radial velocity that our galaxy. Its vertical velocity was less than that of our galaxy. We were moving away from D at the time D emitted the light we are just detecting, with the observed red shift as a result.

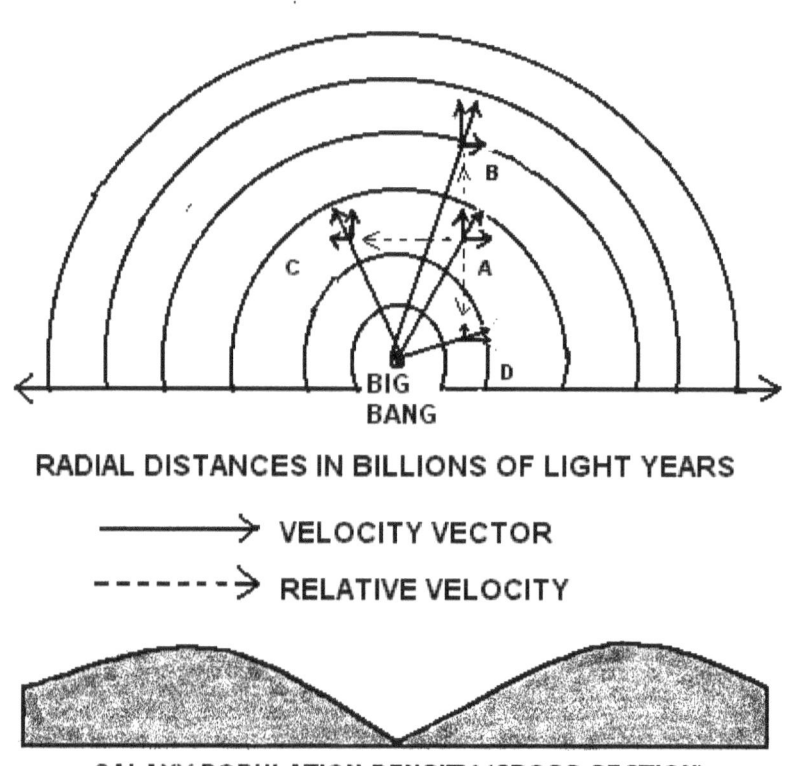

RADIAL DISTANCES IN BILLIONS OF LIGHT YEARS

———————→ VELOCITY VECTOR

- - - - - - → RELATIVE VELOCITY

GALAXY POPULATION DENSITY (CROSS SECTION)

Figure 6

Another possibility is to consider a galaxy that started with about the same radial velocities as our galaxy but at different azimuthal angles. This is shown in the above figure as the galaxy at point C. This galaxy and ours are both moving away from each other. The red shift results from the sum of the horizontal velocities. This red shift would be considerably greater than for the other two cases, even though all are at the same distance from us.

This could account for the almost two to one variations in the measurements of the Hubble Coefficient by different observers and the various estimates for the age of the universe of from 10 billion to 20 billion years.

I will now propose a hypothesis that could explain a number of things. I believe that it is reasonable to assume that other universes preceded ours. While the course of the expansions and collapses of the prior universes cannot be known precisely, we can assume that they ended with conditions that led to the initiation of the next universe. It is likely that there was sufficient matter that collapsed to a central concentration of mass that produced a "Big Bang".

If a model of an expanding and contracting universe were made it might show that by the time an appreciable amount of matter had accumulated at a central point, much of the remaining matter would be relatively close, probably consisting mostly of galaxy-weight black holes. Assuming that the concentration of exotic matter at the central mass had reached a critical density or state, the addition of another massive black hole may have triggered the Big Bang. I believe that the Big Bang was a fact of nature that happened as a naturally occurring event.   Also, if something happened once, then the conditions that led to it can happen again.

When the detonation occurred and the expanding plasma reached the point of atomic particle formation, they could have started responding to the mass of the surrounding black holes.

Some of the high velocity plasma would have impacted or been trapped into orbits around the black holes, perhaps producing what we detect as quasars. The disruptions in the outward flow would have resulted in variations of density and velocity. This could account for the eventual clusters of galaxies and the voids between galactic clusters.

The momentum of the expanding plasma would interact with the momentum of inward moving black holes. The results could be a slowing of the inward motion or a reversal of the motion of the black holes. The massive black holes would probably resume their inward motion with some exhibiting quasar activities.

A black hole with only the mass of a single star would be capable of attracting a concentration of expanding gasses but would be carried along with the gasses. This might explain the initiation of galaxies with black holes at their center.

The following Figures 7 and 8 are my concept of how the universe might have looked at this early time and perhaps the present time.

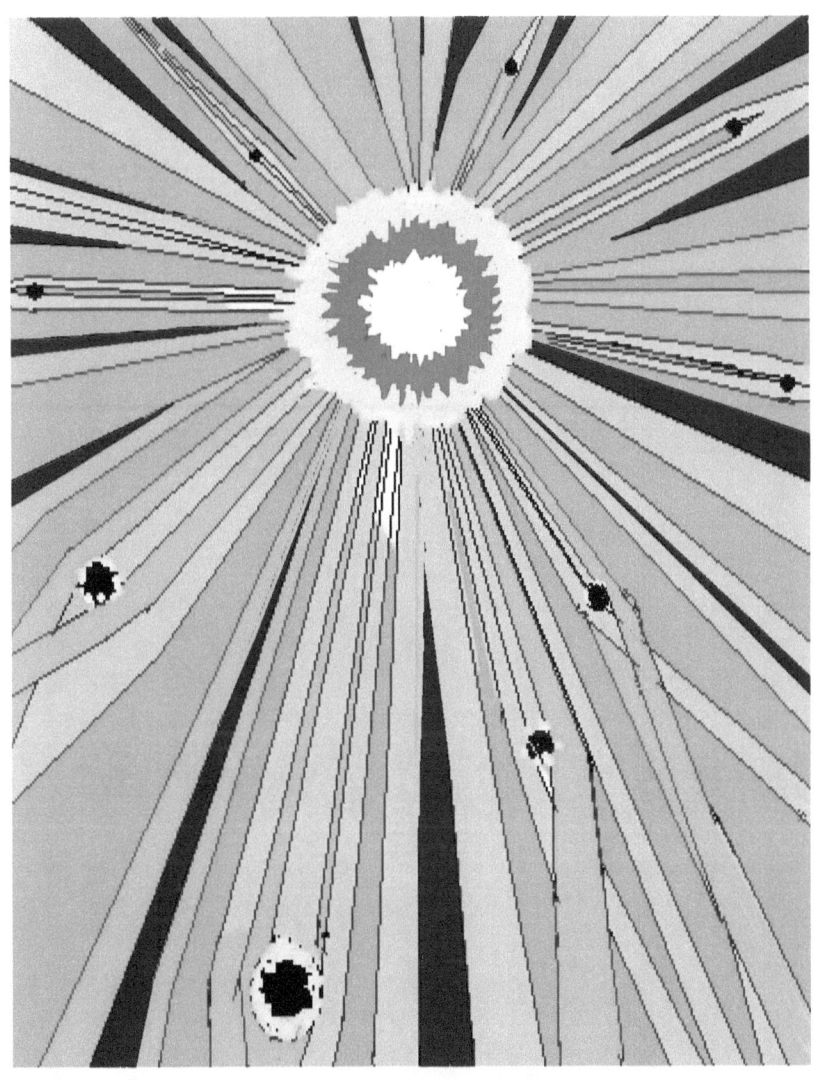

Big Bang and Black Hole Interactions

Figure 7

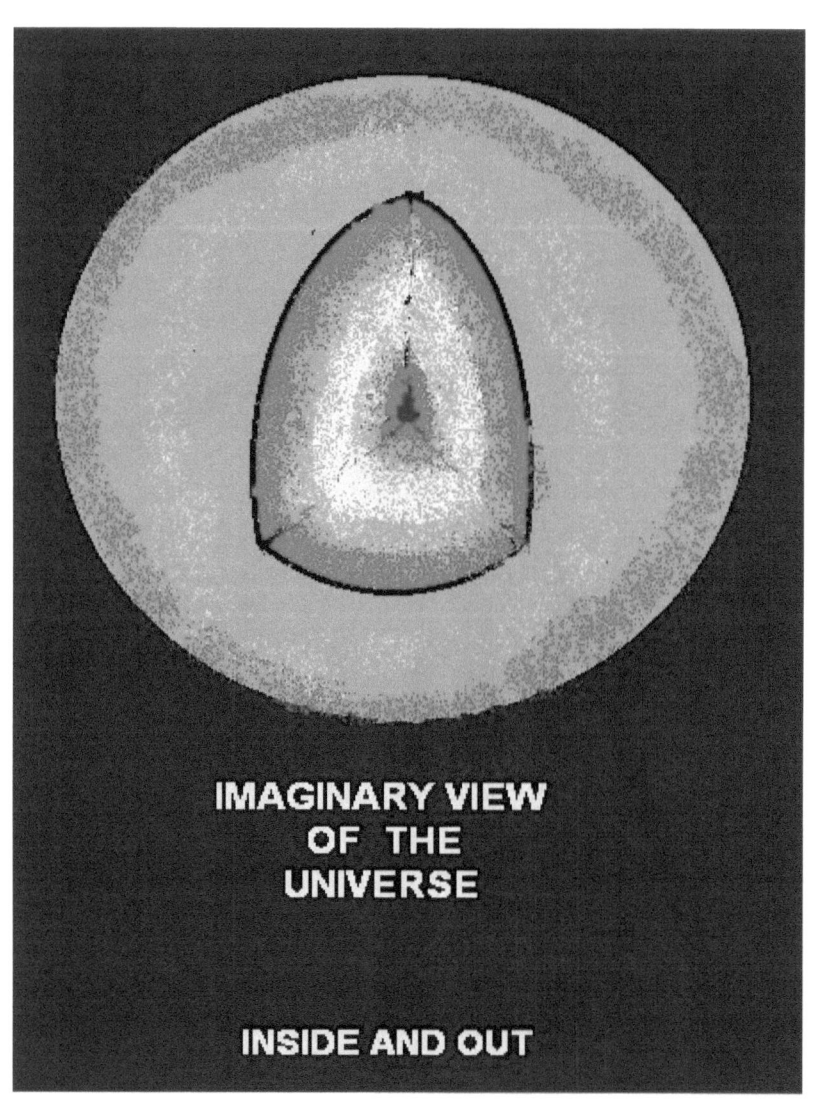

Figure 8

MULTIVERSE

As I considered the report that the apparent increase in red shifts of distant objects was accelerating, the following idea occurred to me. Massive objects lying beyond our expanding universe could affect the outermost objects in the universe. Perhaps other universes surround our universe.

The prefix "uni" in universe comes from the Latin for "one" but in the word "universe" it is considered to mean "the one and only". I believe it is conceivable that our universe may be only one of many simultaneous universes. I would like to call this the "multiverse" consisting of all the universes, since "multi" is the prefix for "many".

I have already written about my concept of sequential universes. However, my ideas about simultaneous universes will need some explaining. I believe that space is endless, infinite in extent, and linearly three-dimensional. I also believe that time is also infinite in extent, without a beginning or end.

Space may curve in the vicinity of matter, according to Einstein's formulation. The more concentrated the matter, the more curvature to adjoining space. It is this curvature, he calculated, that produced the effect called gravity. Apparently this has been observed near the Sun and distant galaxies. We have the Newtonian law of gravity stating that the force of gravity falls off as the inverse square of distance. It stands to reason that with the overall density of our universe being very low, the net curvature is very low.

Assuming that the age of our universe is 20 billion years, at most, then the light emitted at the time of the Big Bang has now reached the surface of a sphere with a radius of 20 billion light years, ignoring the supposed relativity effect on time. The sphere containing any appreciable matter, such as high velocity gasses, probably has a slightly smaller radius. If we consider greater distances, such as 50 or 100 billion light years away, it seems reasonable that the curvature of space will decrease and eventually become linear.

We can now picture our expanding universe as an isolated sphere floating in infinite space. What is wrong with this picture? Why should there be only one unique universe? An infinite space has room for an infinite number of universes. I believe that the existence of any matter and energy at all, means that there can be an infinite amount of matter and energy in existence.

I have pondered on the source of all of this and have come to a conclusion. If one believes that everything was divinely created, then that is the answer. The only alternative that I have come to believe is that all matter and energy always existed. I often come back to thinking; "but where did it all come from originally?" I have to remind myself that it always existed.

If there are other universes surrounding ours, how could they be spaced around ours? With the simplifying assumption that they are all spherical and about the same size, the densest configuration would be for 12 other universes to surround ours.

This would place ours in a somewhat favored position, which I don't think is warranted. They could be located randomly with varying distances and orientations relative to our universe. I think that over a large number of iterations the random configuration would come to have a fairly regular arrangement. This is based on my belief, that over time, they will interact, and create new universes.

A reasonable assumption would be to place our universe in a cubic lattice. This configuration could be repeated indefinitely in all directions with none of the universes in a favored position.

We could picture our universe at the central vertex of the eight surrounding cubic volumes. This closest cubic array has a total of twenty-six other universes. Let us consider the dimension of the side of the cube to be some unit to be determined.

There would be six universes at this closest distance away from ours. This is shown in Figure 9. These six universes are those along the X, Y and Z-axes.

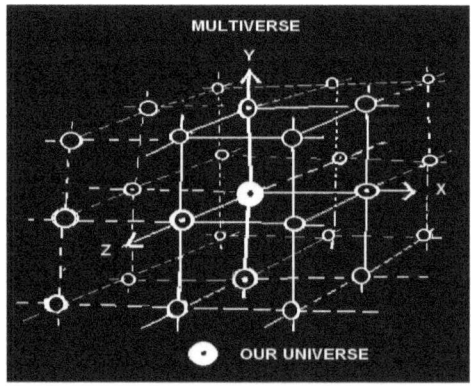

Figure 9

38

There would be twelve more universes centered at a slightly larger distance away, the square root of two times the unit distance. Finally, there are eight more centered at the square root of three times the unit away. The gravitational effect of these surrounding universes should be calculable. At the center of our universe they will produce equal and opposing fields from each opposing pair of universes. This effect could represent what has been proposed as the Higgs field.

As an outermost galaxy in our expanding universe gets closer to one of the other universes and farther away from the opposite universe, it may be effected by a difference in the gravitational effects, and undergo acceleration toward the closer, adjacent universe.

From the acceleration that has been observed, an estimate of that unit of distance between universes might be attained.

Although the surrounding universes may be at different stages of expansion they eventually could reach a point were two or more of them start to overlap. This might produce a situation where in that region the concentration of mass would reach a larger density and start to compress toward a new "Big Bang". Obviously this event would occur in the regions between groups of universes. In this way, universes that are expanding towards oblivion can give birth to new universes by interacting with one another.

That all the universes would begin at about the same time is based on the following assumption. Any expanding universe would spread an eighth of its total mass into each of the adjacent lattice cube volumes.

Each of the universes at the eight corners of each lattice cube would do the same. The largest concentrations of mass would eventually occur in the center of each cube. This is where the total mass would again reach the critical value. Figure 10 shows eight universes expanding toward each other.

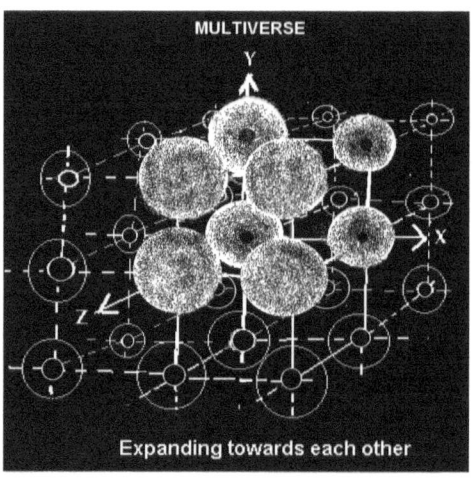

Figure 10

GRAVITY

I have been mulling over how to handle gravity. Newton's approach to gravity is appealing for its simplicity but does not address what causes gravity. It only shows how the effects of gravity can be calculated. He assumed that gravity is a force of attraction acting between material bodies.

Einstein's mathematical approach lead to the concept of curved space-time causing the effects of gravity. By manipulating equations involving multiple variables and coefficients he was able to predict known experimental results. His explanation for gravity was that objects followed paths of least resistance in curved space. I cannot visualize empty space being curved. As a result of his equations, some explain the expanding universe merely as space expanding.

My first approach to this problem progressed along the following train of thought. I have been struck by the need to have gravity act over great distances essentially instantaneously. I think that the gravity of the total mass of our universe may be felt far beyond the present radius of the Big Bang radiation, which is about 10 to 20 billion light years.

What if at the time of the Big Bang, when the initial state consisted only of gravitational force, a gravity wave was radiated at a velocity far in excess of the present speed of light? What if gravity is always transmitted faster than light? Would it be an attracting force or a repelling force? In my concept of multiple universes, I assumed that all the universes surrounding ours began at about the same time.

All of these universes would have about the same total mass as each other. This is based on my concept of an accumulation of giant black holes of the densest kind (particles made of the most massive quarks) reaching a critical mass and collapsing into a Big Bang.

The gravity waves from all of these universes would be permeating our universe from all directions. The effect of this would tend to balance out on the large scale within our universe. However, if these gravity waves were repulsive, a large body could tend to shield a smaller nearby object from the waves coming from beyond the larger body. This would make the smaller body move towards the larger.

I learned that one theory of gravity assumed that gravity was transmitted by an intermediary particle called a graviton. It was said to be a mass-less particle traveling at the speed of light and having momentum. If gravity is transmitted by a mass-less particle, then there is no reason that at would be limited at the speed of light.

Now a new approach to gravity occurred to me. I am proposing that gravity is caused by gravitons transferring momentum to the atoms with which they collide.

The greater the number of protons, neutrons and electrons in a body per unit volume, then the more the gravitons will interact with the body. I am assuming that these will be elastic collisions and not reduce the energy of the gravitons significantly. Many gravitons will pass through most material objects without any collisions. Those involved in collisions will be scattered over a range of angles.

Any isolated material body in space would be subjected to equal amounts of this momentum from gravitons coming from all directions of space. Therefore, there would be no net momentum added in any one direction. However, if a larger body were in the vicinity of a smaller body, it would shield it from receiving some of the gravitons coming from the direction beyond the larger body. This is shown in Figure 11.

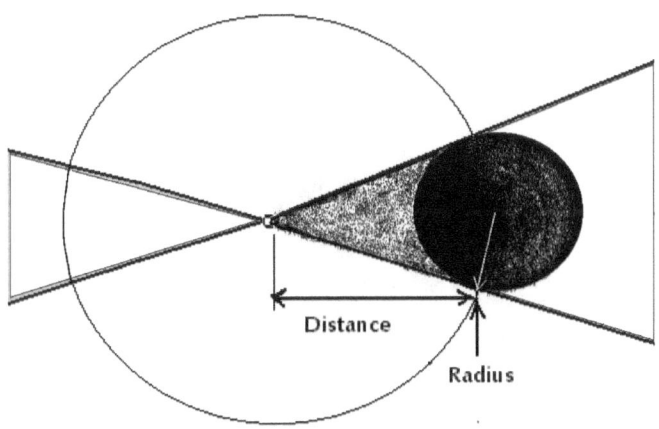

Figure 11

The amount of shielding would depend on the presented area of the larger body and its distance from the smaller one. It also depends on the density of the large body and how many of the gravitons it intercepts and how many it transmits.

Through the same solid angle that the large body presents, but in the opposite direction from the small body, the incoming gravitons are unimpeded.

This results in a net difference in momentum applied to the small body. This will cause it to move towards the larger body with a specific acceleration. The effect of the momentum will depend on the mass of the small body. For a given volume of matter, the greater the density, the greater the mass, and the greater the force applied to the body by the gravitons.

Since force is equal to mass times acceleration, if the ratio of force to mass is constant, then the acceleration is constant.

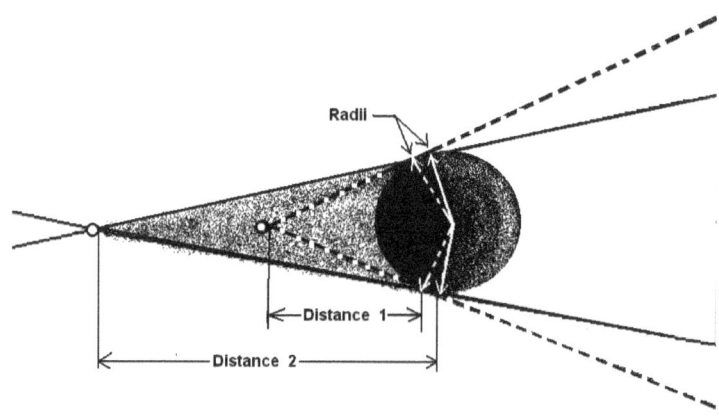

Figure 12

The solid angle presented by the larger body depends on the distance between the bodies. As shown in Figure 12, the solid angle will vary as the inverse square of the distance.

The inverse square relationship can be explained as follows: The presented area of the larger body is approximately pi times its radius squared. This will be a small fraction of the total area of a sphere centered on the smaller body, calculated for the specific distance. The area of this sphere equals four pi times the distance squared.

The fraction of the total radiation intercepted by the larger body is given by the ratio of these expressions. This is the radius of the larger body squared, divided by four times the distance squared. This simplifies to a constant divided by the distance squared.

This applies actually when the presented angle is small. As the smaller body approaches the larger body, this angle will become large and the relationship with distance will deviate from the inverse square.

I do not think that gravitons come from the processes occurring in stars, such as our Sun. If our Sun were the source of these gravitons they would probably outweigh the contribution from all the other stars in our vicinity and the Sun would repel the planets.

There is a very energetic process occurring around us in space. It is the process where black holes ingest matter from adjacent stars.

This has been observed between binary stars with companions. It also occurs in the cores of galaxies where central black holes are swallowing near-by stars. If this is the process that generates gravitons, it is reasonable to expect that they come towards us from all directions. Another energetic process in space is the collapse of stars, such as super nova occurrences.

After some additional thought I have concluded that these somewhat locally occurring events could not account for gravitons throughout our galaxy and universe. If the central black hole of our galaxy were emitting gravitons, it would be repelling the near-by stars.

For this theory to be possible, the source of the gravitons might be external to our universe. This is as described above in the multiple universe idea.

What gravitons are is a good question. They would have to perfuse our space but not have been detected, except for the effect of gravitation they produce. The only theoretical particle that has been postulated that I know of is the Higgs Boson. It has not been detected as yet. There has been some mention of a Higgs Field.

The concept of gravitons traveling at "light speed" has a direct bearing on my concept of photons of light being particles having a finite mass traveling at slightly less than "light speed".

One can visualize a photon traveling and being retarded by the gravitons coming from the direction directly in front. A photon would also be propelled by the gravitons coming from behind it. If the photon reaches the velocity of the gravitons coming from behind, the gravitons will no longer propel the photon. Then the gravitons coming from the front will retard the photon until the balance is restored. This should occur at a velocity just below the velocity of the gravitons.

This process could produce an oscillation of the photon, back and forth along its direction of travel. A more massive, higher energy photon might have a smaller excursion than a less massive photon. This could produce a higher frequency, shorter wave-length for the more massive photon.

It remains to be determined how the oscillation could cause the other motions that I have proposed that would produce the electro-magnetic characteristics of the photon. (See Addendum A.)

I have been reading of attempts to replace the "Inflationary Theory" in the Big Bang. One of the ideas is that at the very beginning of the universe, the velocity of light was far higher than it is now. With my concept of the speed of light being controlled by the speed of gravitons, this would imply that the speed of gravitons, at that time, was much higher than it is presently.

After the collapse of the primeval black hole into pure energy, the first force that is supposed to have emerged is gravity. It has been assumed that this was a force of attraction. But by my assumption, it was a repellant force and would initiate rapid expansion.

With further cooling, the other forces would emerge, and than the primary particles would form. These particles would be affected by the gravitons. Most of the gravitons would be scattered by the particles and remain among them.

As the universe evolved, the greatest portion of the mass of the universe would be in the central part of the distribution at an average radius. Within this region, bodies would be impacted by gravitons coming from every direction. Only in the innermost and the outermost parts of the population would there be an imbalance of the graviton impacts.

Bodies in the innermost region would tend to be retarded and their rate of expansion would be slowed by the predominant direction of the graviton flux being from the central region.

The bodies that are in the farthest out region will also be affected by the graviton flux in an unbalanced way. They will be accelerated by all the gravitons coming from behind them because there is none coming from farther out to retard them. That is, until another universe is approached.

With the new concept for gravity, I have to take another look at the other earlier ideas. I am assuming that black holes are configured similar to neutron stars. There should still be the outside layer of iron atoms. Within that, there should be a layer of neutrons, and inside of that the "charm/strange" particles.

Assuming gravitons impact on a black hole from all directions, what could be the results? The gravitons could be scattered away since they are suppose to be without mass. However, they may enter through the spaces between the surface atoms and make innumerable elastic collisions within the black hole. This would result in more gravitons coming towards the black hole then going away.

Any material body in the vicinity would be accelerated toward the black hole, since few gravitons will come through the black hole to impede them. The approaching bodies would achieve exceedingly high velocities and be annihilated upon impact.

When black holes are in the same vicinity, they will tend to shield each other from gravitons coming from directions beyond them. The gravitons coming from the opposite directions will not be shielded and will impart momentum to the black holes they fall upon. This will cause the black holes to move towards each other, until they merge.

It is evident that black holes would accumulate gravitons. These gravitons will have come from their own universe, from adjacent universes, and even from previous universes that no longer exist. These gravitons will be recycled, liberated when the black holes containing them turn into a Big Bang.

TIME

I have saved the "best" for last, TIME. Since I believe that time had no beginning and will have no end, the question is where to begin?

We all know what time is when we experience it in a straightforward way. The possibility of going back in time has been written about in fiction. Some scientists even suggested that it might be possible to do so. They speak of theoretical "worm-holes" that a space ship could enter and come out at a vastly different place and time. They have also described how particles created by collisions of atoms could be considered as moving backward in time and recombining into the original atoms. I do not believe that any of this is reasonable.

I believe that time is a linear dimension on which our mind places events in logical sequence. We can project backwards and forwards along the time line, using our imagination. This is subjective time. We calibrate our perception of time by how long it takes for our planet to revolve once a day, and to circle the Sun once a year. But there is a real objective time, in which past events occurred in sequence, without a mind being aware of their occurrence.

It has been suggested that time began with the start of our universe. I do not believe that this is true. My number one assumption is the following:

The passage of time had no beginning and will have no end. Time is infinite.

This takes us to a consideration of infinity. In elementary plane geometry we were given the property of two parallel lines that will extend to infinity and still not cross. An alternate form of geometry proposed that parallel lines did cross at infinity. This created the impression that infinity was a far distant place that could be reached. But this is nonsense. Things that extend infinitely never reach an end point. If things have been occurring for an infinitely long time, they never had a beginning.

The theory of relativity says that time varies with the velocity of an object. This idea first arose because of the belief in the "luminous ether." For a hundred years it was assumed that light, as a wave, needed a medium in which to travel through space. This medium was called the "luminous ether."

A problem arose in trying to measure the velocity of light. It had been assumed that the velocity of earth traveling through the ether would produce a change in the speed of the light moving in the direction of the motion of the earth. This was as compared to the speed measured at right angles to this direction.

Many attempts were made to detect this change in velocity, using instruments with ever increasing accuracy. No change in the velocity could be measured.

In a last effort to retain the concept of ether as the medium for electro-magnetic radiation, a theory was proposed that the units of length and time, used to measure the velocity of light, must change as a function of the velocity of light. The equations that described this effect are called Lorentz transformations.

Einstein used these equations in his theory of relativity. He applied them to all matter, except for light. He defined the velocity of light as constant. His equations have time slowing with increased velocity. He also derived equations for the relationship of mass to velocity and the equivalence of mass and energy.

The assumed time dependence on velocity has lead to many science fiction stories. The idea of a person traveling at near light speed and aging much more slowly than someone remaining behind has been intriguing. I believe that this cannot happen, for a number of reasons. We know of no practical way to accelerate a living person to a velocity where a significant change in mass and time would be predicted. Even if one could reach such a velocity, the physical limitations of a human body would not tolerate such changes. Muscles could not cope with a ten-ton body.

An apparent increase in mass with an increase in velocity could be explained by the effect of graviton interactions. Such an effect has been seen using particle accelerators. However, any apparent change in time, as measured with mechanical or even electronic clocks, could be due to the effects of gravitons on the clock mechanisms or atomic particles used for measuring time.

I believe that time is something sensed by the human mind. It is a subjective sensation that can be felt as fast or slow without reference to a clock. In physics, time should only be considered as the ratio of a distance relative to the velocity used in traversing the distance.

I don't believe that time should be treated as an independent variable, as it is in the theory of relativity.

The above is the crux of my difference with the theory of relativity. Relativity refers things to the point of view of observers. I believe that reality does not require observers for actual events and relationships to be true. I believe there is such a thing as absolute velocity in relation to a point in space. I am sure there are simultaneous occurrences throughout the universe, even though no observer can verify them.

BEFORE THE BEGINNING OF TIME

Lying awake some dark night
Try to imagine what it was like
When all of space was empty and black
Before the beginning of time

It takes a while to reach the state
Where darkness is an endless void
Stretching forever in every direction
Before the beginning of time

But feel the pulsing in your chest
Rising and falling with every breath
Telling time with a metronome heart
Before the beginning of time

Imagine being a bodiless mind
Conscious of only empty space
Still you will sense something pass
Before the beginning of time

Thinking one thought after another
Is somewhat like a ticking clock
You cannot think of timeless time
Before the beginning of time

Lenard Metzger
Circa 1986

LIFE

At the end of these writings I realized that I had omitted a very important item, life. No matter how wonderful the cosmos proves to be, if there were no one here to appreciate its beauty, it would be an empty process. Religion has an explanation for the start of life that satisfies many. Those who question that answer are still doubtful about some of the alternatives. It hardly seems possible that the complex form of life found on our world has had enough time to develop since the Earth became habitable. One suggestion has been made that this world was seeded from space. Then the question becomes, where did these seeds come from and how were they made?

It occurs to me that if beings were living at a time when their universe was dying, they might develop such seeds and disperse them into space. They could hope that their universe would become part of another universe and that some of the seeds would survive the Big Bang. If beings in some other dying universe did the same, also contributing seeds to the new universe, then many kinds of life could arise, with the possibility of something like cross-pollination.

We can assume that such advanced races will have perfected genetic engineering. Their seeds would probably incorporate many alternative genomes, programmed to allow for a vast range of environmental conditions.

The most likely vehicles for transporting such seeds would be on small bodies of ice, to provide the water necessary for the seeds to develop. They would also include the initial nutriments. These objects would be released in vast quantities.

Based on the animal population of earth, these progenitor races probably were mammalian, reptilian and insect-like types.

CONCLUSIONS

With all due respect to Albert Einstein, I have presented my ideas of a worldview that may not be as mathematically elegant as his, but hopefully they will be more readily visualized. I see things differently then those who use his equations unquestionably as the starting point for their extrapolations. My ideas may not all prove to be correct, but they have the beauty of simplicity and being somewhat understandable.

Einstein was able to develop his Special Relativity in less than a year. It took him ten years to put together his General Relativity, to tie gravity in with his first approach. He did this with some borrowing from others. He then spent the rest of his life trying, unsuccessfully, to truly unify gravity with the other forces of nature.

In summarizing the ideas I presented here: I believe that the three dimensional view of objects moving out into a universe of non-expanding space is useful to overcome the dead end of the expanding space model. This model predicts that every other body in space will eventually disappear from our view, moving away at a speed faster than light.

I am very sure that our universe is a naturally occurring process that must have a means of reproducing itself. That is why I feel that the idea of multiple universes surrounding ours is reasonable.

The idea that photons have mass is not really too new. It is only the idea that they are limited to a velocity determined by gravitons that is new.

I must say that I consider the idea of gravitons, acting as a repellant force that imparts momentum to objects and produces the action of gravity, is the most significant of my ideas presented here. This also may be the most controversial.

I believe the concept of black holes being knowable as objects that are produced by a natural extension of known processes is a reasonable one. This ties together a progression; from a large star mass black hole, to a galaxy mass black hole, to an universe mass black hole, and to the Big Bang. It seems that the graviton idea is compatible with black holes, except for the mythical effect on time at the so-called event horizon.

I am the most uncomfortable about time being included in equations for explaining the effects of velocity on an object, such as changing its size and mass.

In Addenda A and B, I added my ideas about light photons and radio photons (rhotons). I believe they have the mass of very small bits of the electrons emitting them.

I also, derived a concept of how electrons and nuclei interact and establish electron orbits in atoms. See Addendum C. This led to a unique idea about the structures of the nuclei in all the elements, which I describe in Addendum D. These ideas take us to the very small end of the cosmos.

CRACKPOT IDEA

The following is the only idea that I am presenting that may truly be a crackpot idea. I am assuming that gravity is really caused by gravitons imparting momentum to atoms.

In that case, what would be nice to develop is a composite material or device that acts like an electronic rectifier or diode. A rectifier allows current to flow through it in one direction but not in the other. With such a device applied to gravitons, one could actually get an anti-gravity effect.

From one direction, most of the gravitons would pass through without imparting much momentum. From the other direction, they would impart momentum. This would move the device in the direction of the applied momentum.

In the following diagram (Figure 13) I have suggested how these devices could be used to propel a space ship. I have assumed that they will be in the form of flat panels that can be moved over a range of orientations independently. If the reflective side of the panels were facing toward the rear of the ship, they will be propelled forward. Four of these devices are shown as the dark objects in the bottom view.

After the ship has accelerated to the desired velocity, the devices can be rotated to the axial position, as shown by the one dark device in the side view. In the neutral case, every other panel will have a reversed orientation.

When it is desired to slow the ship down, devices can be rotated so that the reflecting surfaces are facing forward. To stay at any velocity, every other panel can face forward and the others can face the rear.

Another interesting feature would be that, when traveling at the desired velocity, the devices can be moved to the axial position, all facing in the same radial orientation. In this case they will respond to the gravitons coming from all the radial directions and cause the ship to spin around its fore and aft axis. This will result in centrifugal force providing weight to the crew in the area around the circumference of the ship. To stop adding spin, the panels can be alternated in their orientation so that every other one will respond to opposite momentum.

I have shown a central control area with a forward view. It can be arranged to have this central area free to counter-rotate relative to the circumference if it is desired. Many different combinations of the orientation of the panels are available to provide all the necessary control functions for the ship motion.

One last thought; if the ship were allowed to accelerate at a comfortable one g (the force of Earth's gravity) for about a year, it might reach a speed limited by the velocity of the gravitons. Wow, that would be speed of light travel!

CRACK POT IDEA

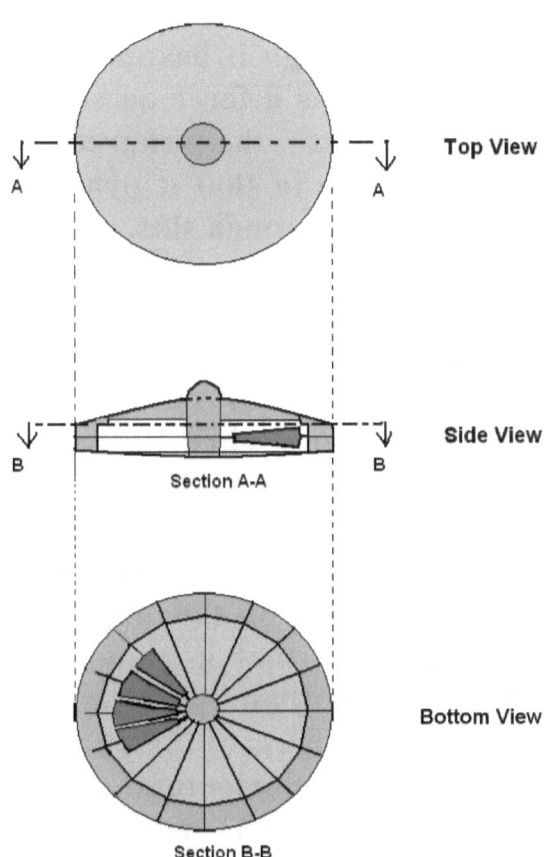

Top View

Side View

Section A-A

Bottom View

Section B-B

Figure 13

ADDENDUM A. LIGHT

The nature of light has been controversial for a long time. A photon of light is considered to be without mass, but acts as though it has mass, since it exhibits momentum. It imparts a force on surfaces it impacts. Light also exhibits the characteristics of an electromagnetic wave in that it produces interference pattern when passed through slits.

The fact that a photon of light has momentum suggests to me that there may be an extremely small amount of mass in a photon. The energy of a photon and the equivalent mass of that energy can be calculated. Perhaps an excited electron gives off a tiny amount of its mass when it emits a photon.

Theory states that for a finite mass to achieve the velocity of light it would take an infinite amount of energy and the mass of the object would become infinite. But suppose that a photon is moving at a velocity slightly less than the ultimate speed of the theory. The effective mass of the photon would be that which accounts for its momentum.

To derive the actual parameters of a photon, start with the equation for a quantum of light, called Planck's Law. The energy of the photon, E is given by the following equation:

$E = hn$ (joules)

h is Planck's constant.

$h = 6.626 \times 10^{-34}$ (joule-seconds)

n is the frequency of the radiation in cycles per second, given by the equation: $n = c/L$

c is the speed of light (c = 3 x 10^8 meters per second)
L is the wavelength of the light in meters.
Therefore:
(Equation 1)　　　$E = hc / L$ (joules)

Assume that the mass of the photon can be determined by using Einstein's equation for the energy:
(Equation 2)　　　$E = mc^2$. (m is mass in kg)

Combining this with Equation 1 gives the following:
$$mc^2 = hc/L$$
Therefore:
(Equation 3)　　　$m = h/cL = (2.2 \times 10^{-42})/L$ (kg)

For green light, $L = 5 \times 10^{-7}$ (meters). Therefore, the mass of a green photon is approximately 4×10^{-36} (kg).

By a different calculation the mass of an electron can be shown to be about 10^{-30} (kg). Therefore, the mass of a green photon is about four millionth of the mass of an electron. It seems reasonable that an excited atom could throw off such a small portion of an electron mass and charge. Traveling at nearly the limiting speed of matter, it will have obtained a relativistic mass that accounts for the momentum of the photon.

To explain the electro-magnetic characteristic of the photon, assume the following: In addition to its forward motion, the tiny charged particle vibrates in a plane transverse to he direction of propagation.

In doing this, the moving charge will induce a magnetic field that will tend to stop its transverse motion at the end of its excursion.

The magnetic field will then start to collapse and cause the charged particle to reverse its sideways motion. As the magnetic field reaches zero, the particle will have reached its maximum transverse velocity and start increasing the magnetic field. This will be of the opposite polarity, which should cause the particle to slow down and stop at the other end of its excursion. Thus, the cyclic transfer of energy between the charge and the magnetic field will continue. This is shown in Figure 14.

It is known that a photon with a higher frequency of oscillation has greater energy content. This suggests that a higher energy photon, with a larger mass and charge, will vibrate faster, with a smaller excursion. It seems possible that the direction of the vibration could rotate around the direction of travel. This would account for circularly polarized light.

A characteristic of electrons that should be considered is that of electron spin. The spin of an electron around an axis through its center produces a circulating charge that causes a magnetic field that emerges at one end of the axis and returns at the other end.

If the small fraction of an electron that is expelled as a photon retains a portion of the spin, this could account for the magnetic portion of the electro-magnetic field of a photon. Its electric field should be oriented transverse to the magnetic field. This is shown in Figure 15.

The fact that the polarity of the electro-magnetic wave reverses every half wave- length that the radiation travels could be accounted for by the particle axis rotating in a plane transverse to the direction of motion.

Another possibility is that the axis of spin could be oriented along the direction of motion. If the axis of spin were to rotate in the plane through the direction of motion, the wave front of the magnetic field would change polarity every half turn and go through a null at the time of reversal.

If the electric field were at a maximum at that instant, when the axis of spin is pointing along the direction of motion, the required alternating maximums of the electro-magnetic field might result.

I have described several possible modes of the photon particle motion to result in an electromagnetic field. I have recently learned that the value of spin assigned to an electron is ½. The photon has been assigned spins of 0 and ½. This might mean that modes I described that assumed spin and no spin might both apply. Figure 14 is a representation of the zero spin possibility.

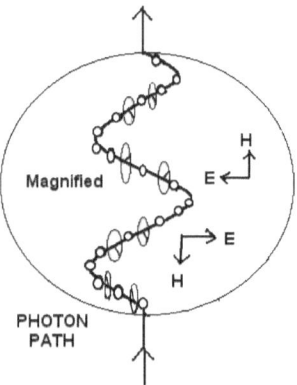

Figure 14

Figure 15 shows how the spinning charge might produce the electro- magnetic fields. The direction of the charged photon path, in this figure, can be either across the page or into the page. A quarter turn of rotation is shown.

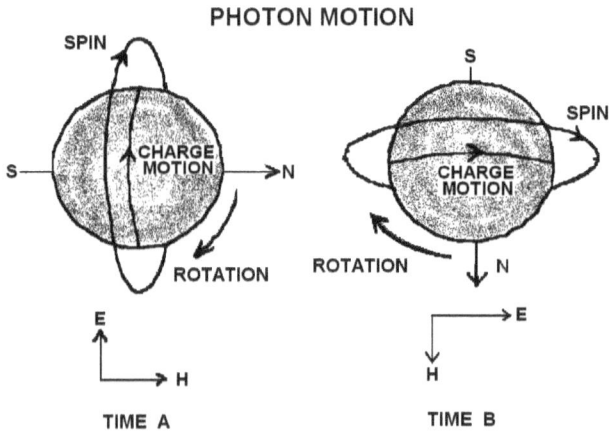

Figure 15

LIGHT DELIGHT

Bright white sunlight
Strikes the Earth's atmosphere
Ultra violet light
Triggers aurora borealis
Sky blue light
Refracts from the ozone layer
Sea green light
Filters down to sunken ships
Rusty red light
Reflects off of Mars
Razor sharp laser light
Bounces off the Moon
Billion-year-old quasar light
Red shifts with age
Invisible light
Stays inside a black hole

Lenard Metzger
Circa 1986

ADDENDUM B. RADIO

It has been generally accepted that all electro-magnetic waves are the same. They are suppose to be described by the Maxwell equations equally well for radio waves and light rays. I have begun to wonder how my concept about photons can be compatible with what I know about radio transmission and reception.

When an alternating voltage signal is applied to a dipole antenna, shown in Figure 16, the following events are said to occur. I will assume a typical FM broadcast frequency of one hundred megahertz. A half wave dipole would be about one and a half meters long.

A sine wave carrier is applied to the middle of the antenna as shown. At the maximum of the applied power the electron flow from one side of the dipole to the other will reach its peak value. A maximum magnetic field will be developed circling the current flow. This field will move away from the dipole at the speed of light.

As the applied power decreases and goes through zero, the electron current flow stops and no magnetic field is formed. At that instant the maximum number of electrons have accumulated at one end of the dipole and the corresponding number of positive charges have appeared at the other end. That is where the electrons have been depleted by the current flow.

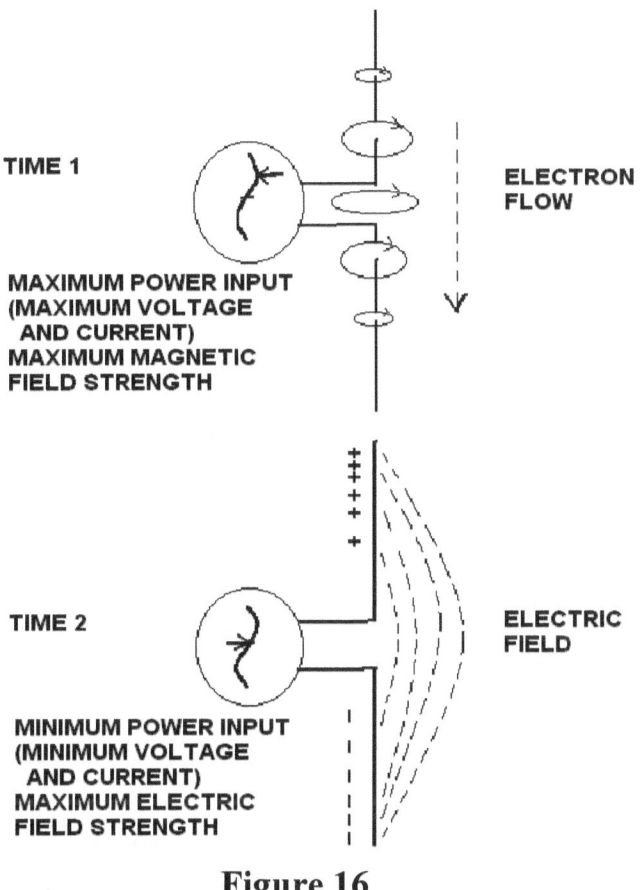

TIME 1

ELECTRON
FLOW

MAXIMUM POWER INPUT
(MAXIMUM VOLTAGE
 AND CURRENT)
MAXIMUM MAGNETIC
FIELD STRENGTH

TIME 2

ELECTRIC
FIELD

MINIMUM POWER INPUT
(MINIMUM VOLTAGE
 AND CURRENT)
MAXIMUM ELECTRIC
FIELD STRENGTH

Figure 16

As shown above, a shell of electric field lines suggests a connection from one side of the dipole to the other. This electric field also moves away from the antenna at the speed of light. It is obvious that in the next quarter of the cycle of voltage the electron flow will reach a maximum value going in the opposite direction. This will develop a magnet field circling in the opposite direction. This field will move off following the electric field, etcetera.

As I mentioned before, electric power is put into the antenna and power is transmitted. The antenna is made of a good electrical conductor so that not much power is wasted in the electron flow, back and forth. It has been assumed that the excess input power is carried off in the electro-magnetic fields. But what if there is the equivalent of photons that carry the power? What if the electric field lines consist of a line of tiny bits of the electrons emitted as the electrons move along the antenna? The bits would have a small motion along the field line. But their primary motion would be away from the antenna "at the speed of light."

There would be an elemental magnetic field ahead of them and another magnetic field of the opposite polarity developing behind them. After that magnetic field would come more tiny bits of electrons with a slight motion in the opposite direction from that of the first particles.

In Figure 17 there are several of the cells, of the many, that form along the length of the antenna. These would interact in a complex interlocking circular motion while traveling outward.

Here I go inventing a new term. If there are radio photons, let us call them "rhotons." They consist of two charge bits per wavelength.

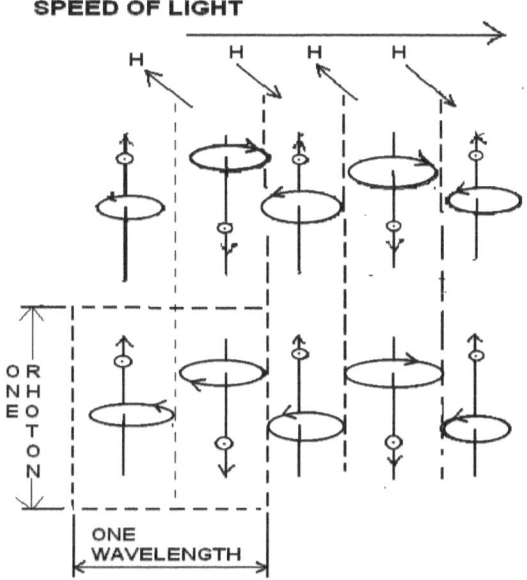

Figure 17

The ovals represent the circulating magnetic flux. The electric charges move only slightly sideways, as they are moving outward at the speed of light. The pairs of charges are linked by their magnetic flux, as they tend to move tangentially past each other. The fluxes add between pairs and produce the alternating magnetic fields of the radiation (indicated by H).

The power from this antenna is radiated in a pattern shown in Figure 18. This is a vertical cross section of the field strength of the indicated dipole. This pattern is a toroid but has also been called doughnut shaped. Looking down at it would show a circular pattern.

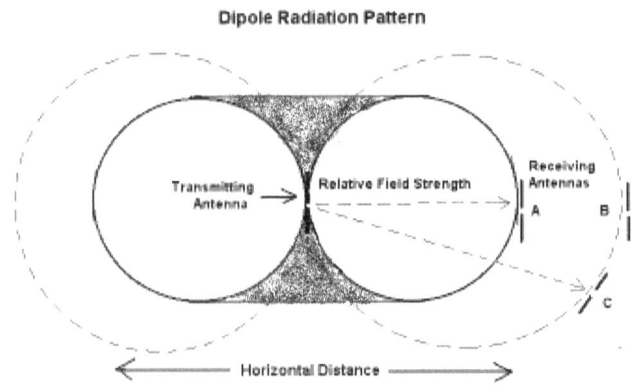

Figure 18

Three different receiving dipoles are shown at different distances and orientation relative to the transmitting dipole. Receivers at A and B are along the line of maximum field strength. B, being farther away than A, is at a point of lower strength. The receiver at C is off the maximum direction and has the same field strength as at point B, but at a closer distance to the transmitting antenna.

It is known that the field strength falls off with distance. With our assumption of elemental cells of oscillating charge and magnetic flux, how can the reduction in strength be explained? With the increasing area that the power is spread over with increased distance, the number of these cells in a unit area will decrease.

If the transmitting antenna is emitting radio type photons, what can be the mechanism? It is known that the electron flow on a transmitting antenna is mostly along the surface. This is called the skin effect.

Also, electrons moving in a metallic conductor, under the influence of an applied voltage potential, migrate by freeing valence electrons and by filling vacant valence sites. Perhaps this low energy transfer process would result in low energy quanta emitted from the surface atoms.

The common metals used in making antennas can be those such as steel or copper. They could be coated with silver. The atomic structure of these metals have from three to four completed electron shells surrounding the nuclei, with only one or two valence electrons in the last incomplete shell. The positive electric field of a nucleus is largely shielded from an incoming electron. So an electron interacts with the atom with a small velocity and a little energy exchange.

If the rhotons of the radio wave are comparable to photons, their characteristics should be calculable. Assume that Planck's equation,
$E = hn$, can be used to determine their energy. The 100 MH FM signal ($n=10^8$), would have an energy per rhoton of 6.6×10^{-26} joule.

Assume the transmitted signal power is 100 watts. That is 100 joules per second. One cycle of the broadcast signal takes 10^{-8} seconds. During one cycle of transmission, the amount of energy transmitted is 10^{-6} joules. The number of rhotons created in this period of time is 10^{-6} divided by 6.6×10^{-26}. This comes out to be 1.5×10^{19} rhotons.

To see how this output number relates to the power input to the antenna, consider that the radiation resistance of a dipole is about 70 ohms. The power input equals $I^2 \times R$, where I is the input current in amperes. Solving for this current: gives the current as I=1.2 amperes. (An ampere is a coulomb per second.) A coulomb is a quantity of electric charge equivalent to 6.28×10^{18} electrons. Therefore, the input to the antenna is about 7×10^{18} electrons per second. During one cycle of input this is 7×10^{10} electrons. Dividing the above number of rhotons per cycle by this number of input electrons per cycle comes out to 2×10^8 rhotons per electron. This seemed like a lot to ask of each electron.

We will use the equation derived for the mass of a photon (Addenda A, Equation 3, m=h/cL) and apply it to a rhoton. In this case L=3 meters. Therefore, the mass of the rhoton would be 7×10^{-43} kg. Comparing this to the mass of an electron (10^{-30}kg), we get the ratio of 10^{12} rhotons per electron. The number of rhotons taken per electron is shown above as 2×10^8. The ratio of these two numbers is 2×10^{-4}. So the mass of the rhotons taken from an electron is only two ten-thousandth of the electron mass.

To get an idea of the density of rhotons in the radiated wave front, I made some simplifying assumptions. Referring to Figure 18 showing the dipole radiation pattern, assume that point A is at a distance of one wavelength, 3 meters. The circumference of this vertical circle is about 10 meters. Most of the radiation will be passing out through the semi-circle, 5 meters around.

The diameter of the horizontal circle will average between 3 and 6 meters. I will assume 4.5 meters. Therefore, the circumference of this circle is about 15 meters. The total area that the radiation will pass through is the product of these two lengths, 75 square meters. By inspecting the derivation of the above area it is obvious that the area will increase by the square of the distance. Therefore the rhoton density will fall off as the inverse square of the distance.

The rhoton density, at any specific distance, will be the number of rhotons transmitted per cycle, divided by the area they pass through, at that distance. As an example; the density at the distance of 3 meters would be 2×10^{17} rhotons per sq. meter. A receiving antenna, placed at that point, would intercept the number of rhotons that would fall on the presented area of the antenna.

If the antenna were 1.5 meters long by .01 meter wide, its area, of .015 square meters, would intercept 3×10^{15} rhotons per cycle. Assume that each of the rhotons, that impinged on the antenna, were to free a valence electron. The total number of electrons that would flow, under the influence of the accompanying magnetic fields, would be less than a thousandth of a coulomb. Hence, less than a milli-ampere of current would be detected. Since, after each half cycle of a rhoton, the direction of its magnetic field reverses, the above current would reverse its direction of flow each half cycle.

ADDENDUM C.

HYDROGEN ATOMS

Going back to the Bohr atom, I wondered at the time, why the electron went into orbit around the proton instead of continuing on and colliding with it. When I learned of the quark theory of the make up of the proton, I began to see why the electron acted as it did. In Figure 19, I show an electron going into orbit around a proton. Obviously, the spacing between the electron and the proton is not to scale.

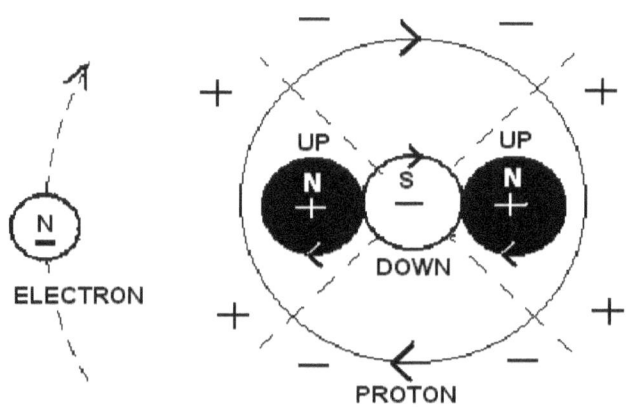

Figure 19

The proton has two Up quarks, each with a +2/3 charge. They would tend to repel each other. It has been stated that something called gluons keep them from flying apart.

The negatively charged Down quark between the Up quarks can tend to hold them together.

The Down quark, with its −1/3 charge, will attract the two Up quarks. The total charge of the proton is the sum of all the quark charges, a positive one.

One trivial point bothered me. Why the −1/3 and 2/3 charges of the quarks? Then I realized that the definition of a unit of charge is completely arbitrary. We could define the Down quark as −1 and the Up quark as +2. Then the proton would be +3 and the electron would be −3.

In addition to their charges, the quarks have spin. This produces a magnetic field around each quark. Assuming that each quark spins in the same direction, the Up quarks will have their positive (north) magnetic poles in the same orientation and the Down quark will have its positive pole in the opposite orientation. This will also cause the magnetism of the Down quark to attract both Up quarks and hold them close.

The entire proton also spins. Its electric and magnetic fields will interact with the electron as they traverse it. The electron spins and has its own electric and magnetic fields. It will be attracted to the Up quarks positive fields but be repelled by the same orientation of its magnetic field with that of the Up quarks.

As the Down quark interacts with the electron, the like electric fields will repel while their oppositely oriented magnetic fields will attract.

At some orbital distance these forces should produce a critical interaction. I believe that it is during this interaction that the photon is produced.

The nature of an electron has not been given much consideration. I believe an electron may have a granular or fluidic composition. Under what could be considered a tidal interaction with a nucleus, an electron may throw off small bits of itself (photons), to achieve an orbit around the nucleus.

Referring to Figure 19, consider the various factors that are involved in the electron and proton interaction. If the velocity of the electron were small compared to the rate of rotation of the proton, successive periods of electric attraction and repulsion would impinge on the electron at the rate of the proton spin. If the velocity of the electron were a significant factor, the lengths of the alternating periods would depend on whether the electron moved in the direction of the rotation of the proton, or in the opposite direction.

To simplify the analysis, assume that the electron does not travel far during a short time interval of interest. The four different electro-magnetic zones of the proton will traverse the electron each revolution of the proton, for a number of revolutions during this time interval.

The electron is also rotating. At some combination of all the factors of the interaction, the electron could be swept by two successive fields of the proton, for every rotation of the electron. The effect of this is shown in Figure 20.

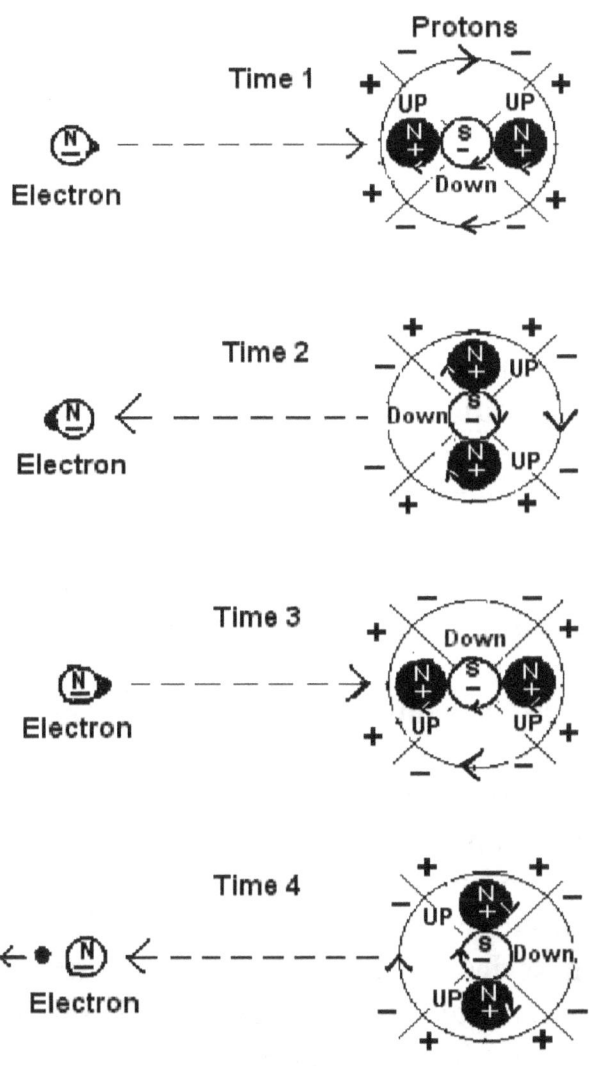

Figure 20

When the Up quark is facing the electron, its positive charge attracts the closest surface of the electron, raising a bulge. By the time that the Down quark is facing the electron, the bulge has been rotated to the opposite side of the electron.

The repulsion of the negative charge of the Down quark, further increases the size of the bulge. This cycle will be repeated until the bulge becomes large enough to separate from the electron and fly off as a photon.

The interaction of the magnetic fields, of the quarks and the electron, could produce motions of the entire electron that are synchronous with the above interaction. This additional motion could enhance the building of the bulge and the separation of the photon.

Assume the photon is launched from the far side of the electron, while the Down quark is repelling the charge, and the magnetic field of the quark is attracting the electron. The contributors to the photons velocity should be the speed and spin of the electron, and the repelling, negative electric field. I am assuming that these cause the photon to achieve the speed of light (almost).

The questions remain, how does this interaction cause the electron to go into orbit around the proton? Does more than one of these bits of charge get thrown off until a stable interaction is achieved? Does a photon consist of a series of charges, perhaps with magnetic fields that alternate, such as a rhoton? If this were the case, the spacing between the bits could determine the wavelength of the photon.

To address the above questions, I will make some additional assumptions. The proton could rotate around an axis that was at right angles to the one in Figure 20. This is shown in Figure 21.

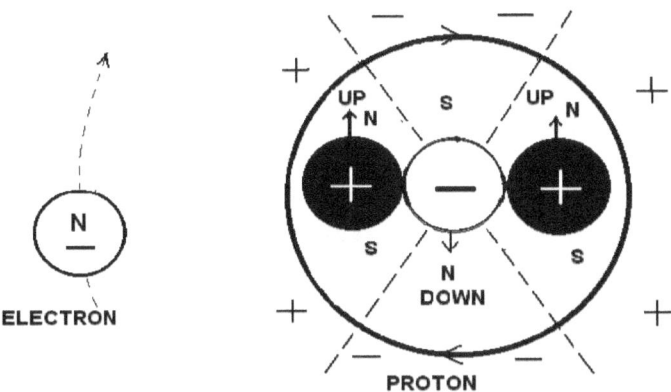

Figure 21

In this case, the magnetic fields of the quarks would not interact with that of the electron. Only the electric fields of the quarks would alternately attract and repel the electron. This case might also tend to produce the bulge, and hence the photon. I think the contribution of the magnetic fields is needed.

I will assume that the spin rates of the electron and the proton do not change during this encounter. The main variables to consider are the velocity of the electron and the distance from the proton to the electron. I will also assume that the direction of the electron motion is the same as the direction of the sweep of the electric fields coming from the proton.

It seems that a stable orbit could come about when the electron is moving at the same velocity as the nucleus fields. The fields would no longer alternate across the electron and would not cause photons. This would mean that the velocity of an electron in one of the larger, outer orbits would be greater than the velocity of one in a smaller, inner orbit.

This is because each electron would have to make a single orbit in the time the proton rotates once. At any other electron velocity or distance, the effect of the alternating fields would depend on their rate of alternation, relative to the spin rate of the electron.

It can be seen in Figure 20, that two rotations of the electron are shown for every single rotation of the proton. Other odd multiples of this electron spin rate could also produce similar alternating interactions. Perhaps photons would occur with six rotations of the electron for each proton rotation, as well as for ten. Such a progression might occur as synchronous rotation is approached.

As a photon is emitted, the momentum imparted to the photon is subtracted from the momentum of the electron. This will tend to slow the electron, and drop it towards a lower orbit. It will go to an orbit where it can again revolve around the proton, in synchronism with the motion of the electric fields of the proton.

I realize that this concept is highly speculative. I understand that other hydrogen spectra are associated with elliptical orbits. It is apparent that the fields of the Down quark would be less intense than those of the Up quarks. These, and other objections can be raised. However, it might be worthwhile to consider this approach to subatomic interactions.

To continue, consider the deuterium atom. Figures 22 and 23 show two possible configurations of the deuterium nucleus.

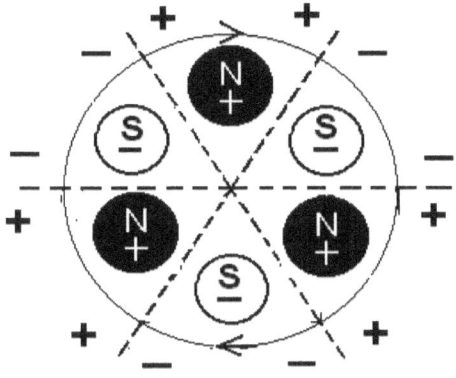

Figure 22

In Figure 22, the configuration would allow alternating interactions, with an approaching electron, from both electric and magnetic fields. In Figure 23 only the electric fields would produce effects on the electron. This is assuming that the electron spin axis is parallel with those of the quarks in Figure 22.

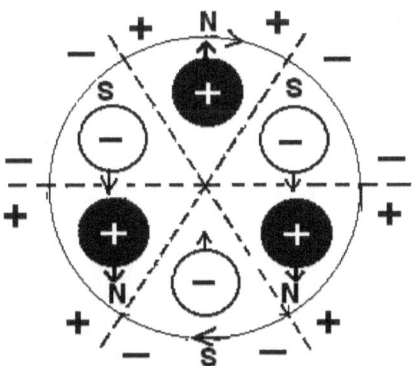

Figure 23

An approaching electron would experience six alternations of the fields for each rotation of the deuterium nucleus. The electron would have to rotate three times, during a single rotation of the nucleus to produce the same kind of interaction as is shown in Figure 20. I assume that odd multiples of this electron spin rate, such as nine and fifteen, would also produce similar effects.

I should repeat, that it is not the actual spin rate of the electron that is different. It is the relationship between the nucleus spin rate and that of the electron that changes. Since the deuterium nucleus has twice the mass of the proton, it is reasonable that it would rotate slower than the proton. Then the required electron distance and velocity, in the two atoms, should be similar when they produce photons.

There are several conditions that could prevent an electron from going into orbit around a nucleus. The electron could approach too fast to be slowed sufficiently. It could be coming on the wrong side of the nucleus, and would have to circle against the sweep of the nuclear fields. The electron could come on the correct side but with the wrong direction of the electron spin polarity.

If an electron approached the nucleus from a direction along the nuclear spin axis, the nuclear fields would not affect it. It would respond to the net electric fields of the nucleus and the surrounding electrons. Such an electron would either be deflected or would continue on to impact the nucleus.

ADDENDUM D.

OTHER ATOMS

I have been giving considerable thought to how the helium atom might work. It would have to be compatible with handling two valence electrons. The electrons would need to have opposite spin axis polarities.

I think that what I show in Figure 24 is a possibility. It is essentially two deuterium atoms, with one flipped over and sandwiched with the other. I assume that this nucleus rotates slower than the deuterium nuclei did before. This could represent a net loss of rotational energy.

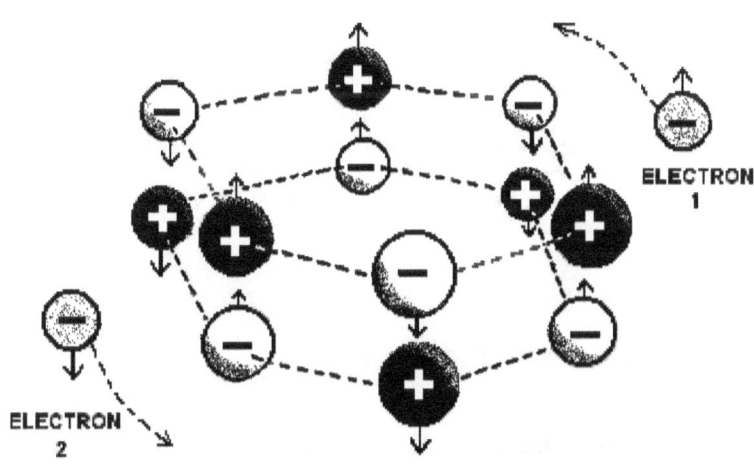

Figure 24

The helium nucleus is rotating in the same direction as the two electrons. One of the electrons will have interacted with one of the deuterium rings and produced one or more photons. The other electron has the opposite spin polarity and will have interacted with the other deuterium ring to produce photons. Both electrons will be in similar orbits, relatively close to the nucleus. The magnetic fields are essential for the correct interactions to occur.

I have also tried to extend this concept to the lithium atom. Since it contains an additional proton and electron, I assumed an additional neutron. Therefore, I added another deuterium ring to the helium atom, with the results shown in Figure 25.

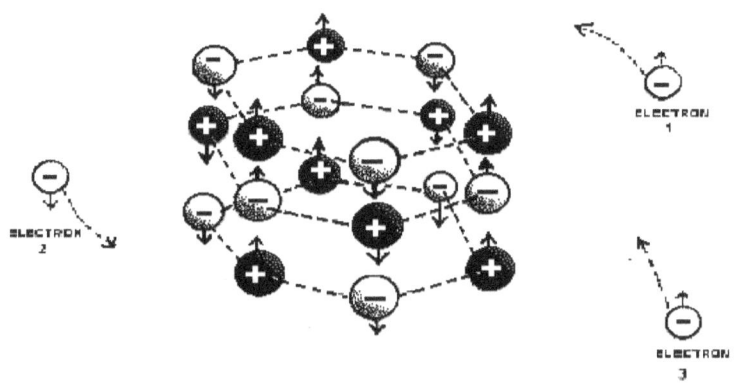

Figure 25

It seems that this configuration of lithium, with three protons and three neutrons, occurs a very small percentage of the time in nature. The majority of lithium nuclei contain a fourth neutron. This extra neutron could be inside the nucleus, as shown in Figure 26. This could add to the stability of the nucleus.

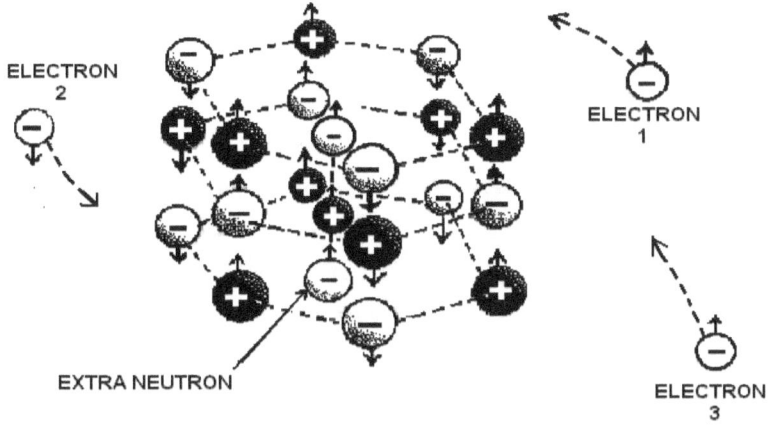

ELECTRON 2

ELECTRON 1

ELECTRON 3

EXTRA NEUTRON

Figure 26

The first two electrons of the lithium atom are in the innermost orbital region, as in the helium atom. The third electron must occupy an orbit farther out, reacting to the electric and magnetic fields that are the result of the additional proton and neutron.

I have included tables of the elements, in Addendum E. These tables give more details about the various atoms.

The next elements, shown on Table 1 are beryllium and boron. Each add a single proton and neutron to the previous atom's nucleus. Of course, they each have one more valence electron than the previous atom.

The following element, carbon, adds only a proton to the number in the boron nucleus. This evens up the number of protons and neutrons to six of each. This is called, carbon twelve. Two other isotopes of carbon exist, having one or two additional neutrons.

The next two elements, nitrogen and oxygen, each add a combination of a single proton and neutron to the previous nucleus.

The heavier elements all have increasing numbers of neutrons, in addition to the number of protons. I feel that there might be a physical picture of how atoms are built.

Based on the above information, I have reached some conclusions and have made the following assumptions:

1. As the atomic nuclei get larger, with more protons and even more neutrons, they must build up a fixed structure that exposes all the protons on the exterior surface of the structure. This provides the mechanism for controlling the orbiting electrons in their proper locations.

2. The protons, with neutrons interposed, are assembled into variously sized rings. This configuration will produce the alternating fields to interact with the orbiting electrons.

3. The surplus neutrons will accumulate within the rings and could assemble into a support and connecting structure for the rings.

4. The pattern of the electron numbers in the orbital shells will have been caused by the pattern of the protons in the associated rings of each element.

5.   The fact that the electrons in each of the allowed orbits are in pairs, suggests that they have oppositely directed, magnetic polarity. I have assumed that the ring structure will also have pairs of rings with oppositely directed, positive quark polarities.

6.   The larger the proton rings of the nucleus become, the larger their orbital electrons` spin rates should become to cause photon emission.

To illustrate these ideas, I have adopted a simplified way of drawing the protons and neutrons in the various sized rings, as rectangles with dark dots for Up quarks and lighter dots for Down quarks. It should be realized that the rings probably assume a circular shape, with alternating positive and negatively charged quarks. In Figure 27 the simplest 2-proton and 3-proton rings are shown in both formats.

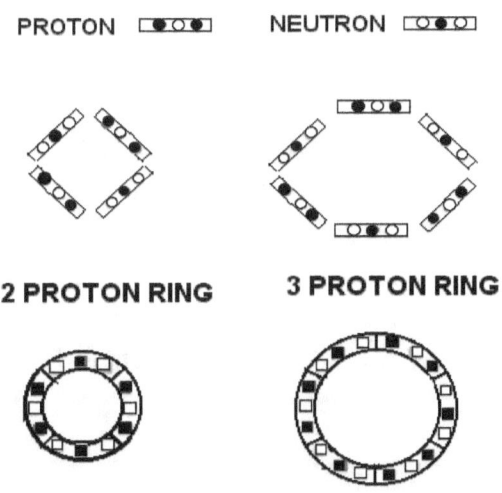

Figure 27

Similarly, in Figure 28, I show the next rings, with increasing numbers of protons.

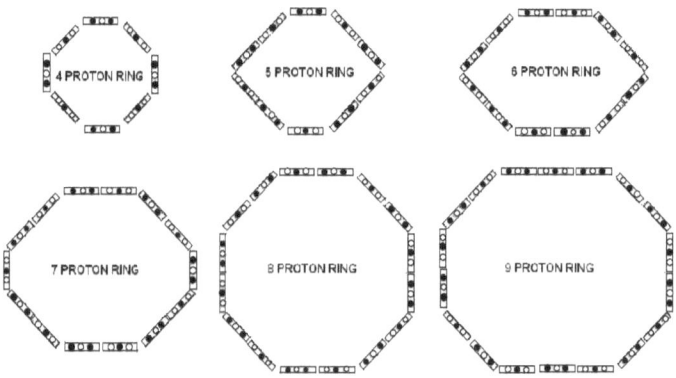

Figure 28

I am assuming that electrons in the closest orbits, controlled by the one proton rings, have a minimum of three times the spin rate of the nucleus when they emit photons. The larger rings should have electrons with spin rates that increase by a factor proportional to the number of protons in each ring. This should give these more distant electrons a stronger magnetic field to better interact with the nucleus and emit photons.

Illustrated in Figure 29, is an assembly of neutrons, having a linear and branching configuration. This could act as the central axis of the nucleus, with the branches connecting to the rings. The negatively charged Down quarks at the ends of the neutrons should be attracted to positively charged Up quarks.

Figure 29

How a ring would fasten to a neutron branch is shown in Figure 30. The center neutron would be part of the axial neutron string.

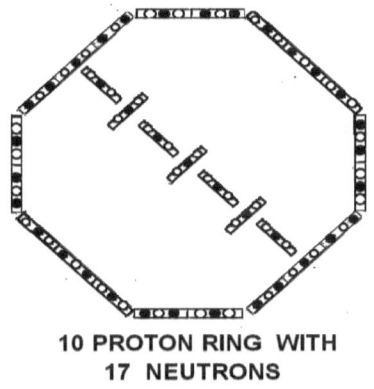

**10 PROTON RING WITH
17 NEUTRONS**

Figure 30

Pairs of rings with the same number of protons, but with opposed magnetic polarities, should hold together the same as the rings in the helium example. (See Figure 24.) But if there is too great a disparity in the proton number, from one ring to the next, some additional neutron bracing might occur, such as is shown in Figure 31.

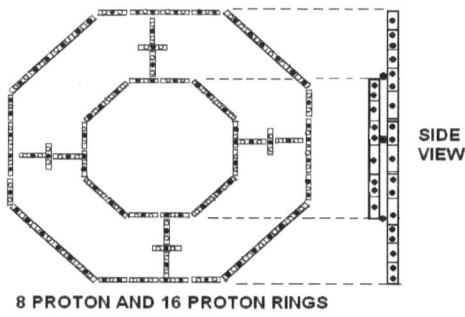

8 PROTON AND 16 PROTON RINGS

Figure 31

In this figure, a 16-proton ring is shown, adjacent to an 8-proton ring. Four linear neutron strings connect the two rings together. It should be noted that the negative quarks at the ends of the neutron strings, connect to positive quarks at each end. I have included a side view of the rings, which I will discuss below.

Another simplifying technique used to help visualize the various atomic structures is to show the side view of a ring. It is shown as a vertical rectangle. The number of protons in each ring is given. This is shown in Figure 32.

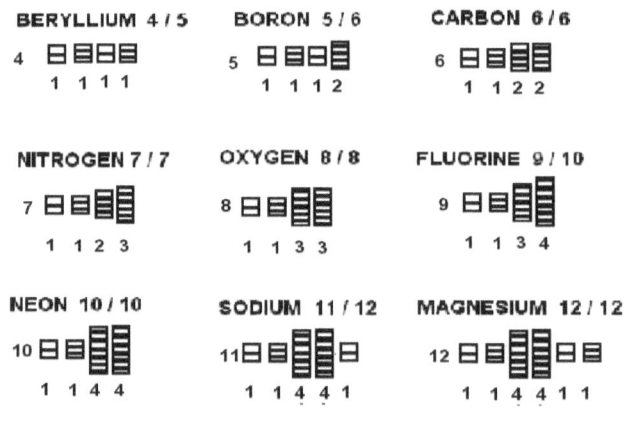

Figure 32

This figure starts with the beryllium nucleus in the upper left corner. This follows the 3-proton nucleus of lithium, as shown in Figure 25. But in Figure 32, the axis has been shift to horizontal instead of vertical.

Beryllium is indicated to have 4 protons and 5 neutrons, and is shown with four 1-proton rings. The neon nucleus is shown with two 1-proton rings and two 4-proton rings. These rings control first and second electron shells that are completely filled. Thereafter, the first four proton rings of all the elements are like this. Since this is such an important interface, I show in Figure 33 how a one and a four-proton ring can be tied together.

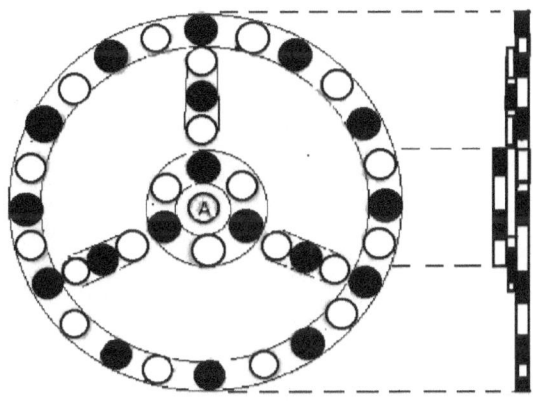

Figure 33

It can be seen that there are six quarks in the one-proton inner ring and twenty-four quarks are in the four-proton outer ring. At most, three neutrons can bridge between corresponding pairs of positive quarks. An "A" indicates an axial quark in the center.

The next nucleus, in Figure 32 is sodium. It has a single proton ring added to the right end, controlling the valence electron in the highest orbit. Magnesium has a second 1-proton ring added. The subsequent elements increase their number of protons in these rings until a total of eight is reached, with two, 4-proton rings. I assume that the following elements will increase the size of their proton rings according to the pattern of electron increases shown in Tables 1 through 3.

For completeness, I am showing some of the heaviest nuclei in Figure 34. These are listed in Table 3.

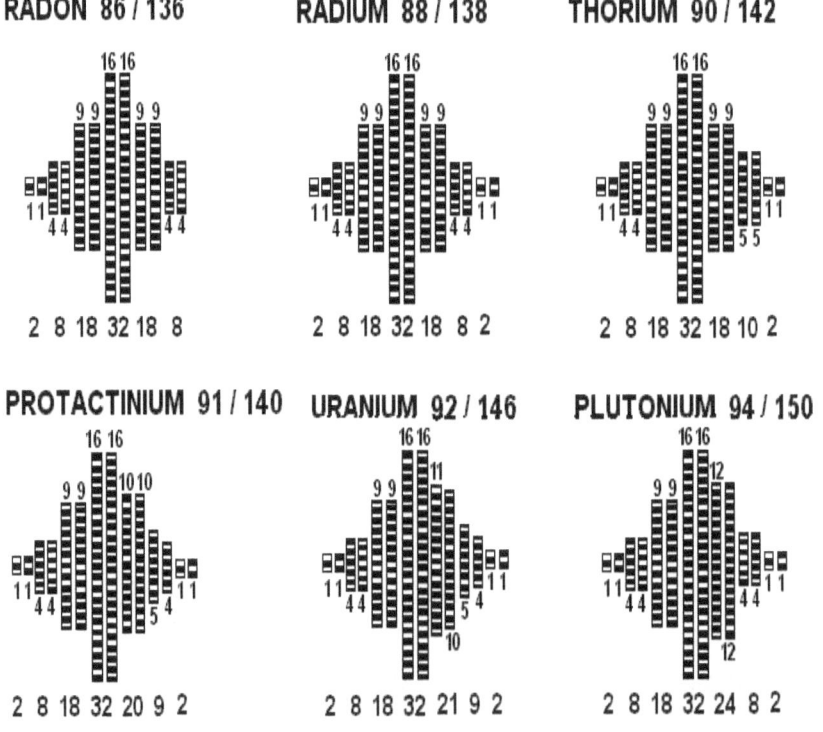

Figure 34

Radon is the heaviest of the inert gasses. It shows the pattern of completed shells, from 2 to 8 to 18 to 32, without a final valence shell. Uranium is the heaviest, naturally occurring element. Plutonium is the familiar, man-made element.

In all of these nuclei one can see how many additional neutrons are included, by noting the second number given. For example, in uranium, with 92 of the neutrons paired with the protons, the remaining 54 neutrons are inside the nucleus.

Note that in all of the above elements the maximum size rings occurs near the middle of the ring structure. They have 32-protons in a pair of 16-proton rings. The following rings get progressively smaller and control electrons in the higher orbits. The electrons for the 16-proton ring should have a spin rate of about 48 times the nucleus spin rate when their photons are emitted.

A one-proton ring to control a valence electron in the highest orbit should have an electron with a spin rate larger than the spin rates of the electrons in the next lower orbit. The greater spin of this electron should give it a larger magnetic field strength to interact with the nucleus.

For one last point of interest, I am including Figure 35. It shows the heaviest man-made nucleus that I have found. It is called unumbium. Smashing zinc and lead nuclei together made unumbium.

The name of this element comes from the Latin for its atomic number: one, one, and two. This is a very short- lived element. Its half- life is about a quarter of a second.

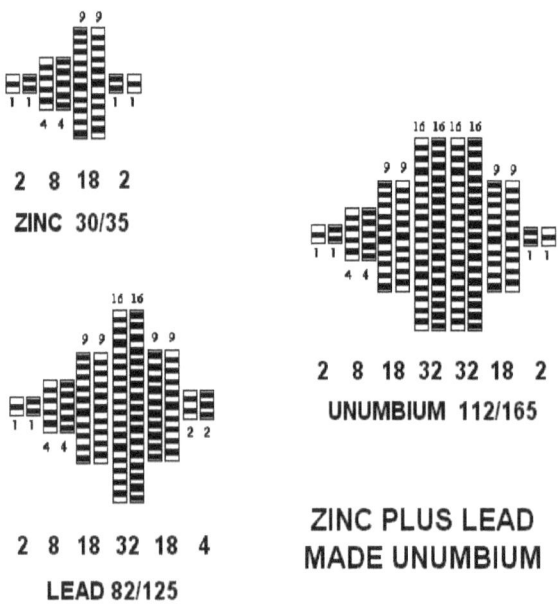

Figure 35

It is interesting to note that the first four ring pairs of the lead nucleus became the front of the unumbium without change. Also, the last two ring pairs of the zinc nucleus became the back of the unumbium. All the other parts of the lead and zinc went into the additional pair of 16-proton rings of the unumbium.

This picture of atoms presents some questions to be considered, such as, "What makes the nucleus spin?" Remember that heavier atoms were built up, in the stars, by adding helium, deuterons and neutrons to lighter atoms. As the heavier, ionized atoms attracted electrons, some of the angular momentum, that the electrons gave up to go into orbit, was added to the angular momentum of the nuclei. This increased the spin of the nuclei.

Another question is, "How do the new electrons know where to go into orbit?" Modifying an atom, which already had its full complements of electrons, made a new ionized atom. An electron approaching the nucleus of this ionized atom, would find that only the ring with the added proton was lacking an electron. The new electron would avoid the other electrons and respond only to this ring, until a stable, synchronized, orbit was achieved.

All the electrons in an atom can be visualized as being in a flat halo, circling the nucleus in unison. Each electron would tend to stay in its own place, in a relatively stationary orbit over its own proton ring.

Any tendency of an electron to deviate from this stability would be corrected by its interaction with the multitude of electro-magnetic fields. The above concept gives a consistent, logical approach to understanding how atoms may be structured.

ADDENDUM E.

TABLE 1

Elements		Atomic Nuclei		Isotope Neutrons	Electrons in Orbital Shells					
		Protons	Neutrons		No.1	No.2	No.3	No.4	No.5	no.6
H	hydrogen	1	0	1 and 2	1					
He	helium	2	2	1	2					
Li	lithium	3	4	3	2	1				
Be	berylium	4	5	6	2	2				
B	boron	5	6	5	2	3				
C	carbon	6	6	7 and 8	2	4				
N	nitrogen	7	7	8	2	5				
O	oxygen	8	8	9 and 10	2	6				
F	fluorine	9	10	9	2	7				
Ne	neon	10	10	11 and 12	2	8				
Na	sodium	11	12	11 and 13	2	8	1			
Mg	magnesium	12	12	13 and 14	2	8	2			
Al	aluminum	13	14	13	2	8	3			
Si	silicon	14	14	15 and 16	2	8	4			
P	phosphorus	15	16	15 and 17	2	8	5			
S	sulfur	16	16	17,18 and 20	2	8	6			
Cl	chlorine	17	18	20	2	8	7			
Ar	argon	18	22	18 and 20	2	8	8			
K	potassium	19	20	22	2	8	8	1		
Ca	calcium	20	20	22 to 28	2	8	8	2		
Sc	scandium	21	24	23 to 28	2	8	9	2		
Ti	titanium	22	26	24 to 28	2	8	10	2		
V	vanadium	23	28	25 to 29	2	8	11	2		
Cr	chromium	24	28	26, 29 and 30	2	8	12	2		
Mn	manganese	25	30	28	2	8	13	2		
Fe	iron	26	30	28, 31 and 32	2	8	14	2		
Co	cobalt	27	32	29 - 34	2	8	15	2		
Ni	nickel	28	31	28 - 37	2	8	16	2		
Cu	copper	29	35	32 - 37	2	8	17	2		
Zn	zinc	30	35	32-42	2	8	18	2		
Ga	gallium	31	39	35-41	2	8	18	3		
Ge	germanium	32	41	34-43	2	8	18	4		
As	arsenic	33	42	38 - 45	2	8	18	5		
Se	selenium	34	45	39 - 47	2	8	18	6		
Br	bromium	35	45	41-50	2	8	18	7		
Kr	krypton	36	48	41-54	2	8	18	8		
Rb	rubidium	37	48	44 - 53	2	8	18	8	1	
Sr	strontium	38	50	47 - 56	2	8	18	8	2	

TABLE 2

	Element	Atomic Nuclei			Electrons in Orbital Shells						
		Protons	Neutrons	Isotope Neutrons	No. 1	No. 2	No. 3	No. 4	No. 5	No. 6	No. 7
Y	yttrium	39	50	47-56	2	8	18	9	2		
Zr	zirconium	40	51	47-58	2	8	18	10	2		
Nb	niobium	41	52	49-56	2	8	18	12	1		
Mo	molybdenum	42	54	49 - 59	2	8	18	13	1		
Tc	technetium	43	55	52 - 57	2	8	18	14	1		
Ru	ruthenium	44	57	52 - 62	2	8	18	15	1		
Rh	rhodium	45	58	56 - 61	2	8	18	16	1		
Pd	palladium	46	60	56 - 64	2	8	18	18			
Ag	silver	47	61	57 - 64	2	8	18	18	1		
Cd	cadmium	48	64	59 - 70	2	8	18	18	2		
In	indium	49	66	63 - 68	2	8	18	18	3		
Sn	tin	50	69	62 - 76	2	8	18	18	4		
Sb	antimony	51	71	67 - 78	2	8	18	18	5		
Te	tellerium	52	76	67 - 82	2	8	18	18	6		
I	iodine	53	74	69 - 83	2	8	18	18	7		
Xe	xenon	54	77	68 - 84	2	8	18	18	8		
Cs	cesium	55	78	74 - 84	2	8	18	18	8	1	
Ba	barium	56	81		2	8	18	18	8	2	
La	lanthium	57	62		2	8	18	18	9	2	
Ce	cerium	58	82		2	8	18	20	8	2	
Pr	praseodymium	59	82		2	8	18	21	8	2	
Nd	neodymium	60	84		2	8	18	22	8	2	
Pm	promethium	61	84		2	8	18	23	8	2	
Sm	samarium	62	88		2	8	18	24	8	2	
Eu	europium	63	89		2	8	18	25	8	2	
Gd	gadolium	64	93	85 - 97	2	8	18	25	9	2	
Tb	terbium	65	94		2	8	18	27	8	2	
Dy	dysprosium	66	97		2	8	18	28	8	2	
Ho	holmium	67	98		2	8	18	29	8	2	
Er	erbium	68	99		2	8	18	30	8	2	
Tm	thulium	69	100		2	8	18	31	8	2	
Yb	ytterbium	70	103	98 - 106	2	8	18	32	8	2	
Lu	lutetium	71	104		2	8	18	32	9	2	
Hf	hafnium	72	107		2	8	18	32	10	2	
Ta	tantalium	73	108		2	8	18	32	11	2	
W	tungsten	74	110		2	8	18	32	12	2	
Re	rhenium	75	111		2	8	18	32	13	2	
Os	osmium	76	114		2	8	18	32	14	2	
Ir	itidium	77	115		2	8	18	32	15	2	
Pt	platinum	78	117		2	8	18	32	17	1	
Au	gold	79	118	115 - 120	2	8	18	32	18	1	
Hg	mercury	80	121		2	8	18	32	18	2	

TABLE 3.

	Elements	Atomic Nuclei		Isotope	Electrons in Orbital Shells						
		Protons	Neutrons	Neutrons	No. 1	No. 2	No. 3	No. 4	No. 5	No. 6	No. 7
Tl	thallium	81	123		2	8	18	32	18	3	
Pb	lead	82	125		2	8	18	32	18	4	
Bi	bismuth	83	125		2	8	18	32	18	5	
Po	polonium	84	125		2	8	18	32	18	6	
At	astatine	85	125		2	8	18	32	18	7	
Rn	radon	86	136	130 -135	2	8	18	32	18	8	
Fr	francium	87	136		2	8	18	32	18	8	1
Ra	radium	88	138	134 - 139	2	8	18	32	18	8	2
Ac	actinium	89	138		2	8	18	32	18	9	2
Th	thorium	90	142		2	8	18	32	18	10	2
Pa	protactinium	91	140		2	8	18	32	20	9	2
U	uranium	92	146	138 - 148	2	8	18	32	21	9	2
Np	neptunium	93	144		2	8	18	32	23	8	2
Pu	plutonium	94	150		2	8	18	32	24	8	2
Am	americium	95	148		2	8	18	32	25	8	2
Cm	curium	96	151		2	8	18	32	25	9	2
Bk	berkelium	97	150		2	8	18	32	26	9	2
Cf	californium	98	153		2	8	18	32	28	8	2
				Half Life							
Es	einsteinium	99	153	1 yr.	2	8	18	32	29	8	2
Fm	fermium	100	157	3 min.	2	8	18	32	30	8	2
Md	mendelevium	101	157	50 sec.	2	8	18	32	31	8	2
No	nobelium	102	157	2 min.	2	8	18	32	32	8	2
Lr	lawrencium	103	159	13 sec.	2	8	18	32	32	9	2
Rf	rutherfordium	104	157	5 sec.	2	8	18	32	32	10	2
Db	dubnium	105	157	4 sec.	2	8	18	32	32	11	2
Sg	seaborgium	106	157	.3 sec.	2	8	18	32	32	12	2
Bh	bohrium	107	157	.1 sec.	2	8	18	32	32	13	2
Hs	hassium	108	157	2 ms.	2	8	18	32	32	14	2
Mt	meiterium	109	157	3 ms.	2	8	18	32	32	15	2
Uun	ununnilium	110	159	10 ms	2	8	18	32	32	17	1
Uuu	unununium	111	161	___	2	8	18	32	32	18	1
Uub	unumbium	112	165	.3sec.	2	8	18	32	32	18	2

(This is from a self-portrait by the author)
Circa 1968

Lenard Metzger was born and raised in Rochester, New York. He has lived there ever since, except for a few years in service and college.

A long time employee of Eastman Kodak Company, he retired in 1983. He then returned to his early interests in art and writing.

www.ingramcontent.com/pod-product-compliance
Lightning Source LLC
Chambersburg PA
CBHW030901180526
45163CB00004B/1661